HYPERFINE INTERACTIONS IN EXCITED NUCLEI

# HYPERFINE INTERACTIONS IN EXCITED NUCLEI

Editors

GVIROL GOLDRING
The Weizmann Institute of Science, Rehovot

and

RAFAEL KALISH
Technion-Israel Institute of Technology, Haifa

VOLUME 3

GORDON AND BREACH SCIENCE PUBLISHERS

New York    London    Paris

Copyright © 1971

GORDON AND BREACH SCIENCE PUBLISHERS INC.
440 Park Avenue South, New York, N.Y. 10016

Library of Congress Catalog Card Number: 78-127883

*Editorial Office for Great Britain*:

Gordon and Breach Science Publishers Ltd.
12 Bloomsbury Way
London W.C.1

*Editorial Office for France*:

Gordon & Breach
7-9 rue Emile Dubois
Paris 14$^e$

All rights reserved. No part of this book may be reproduced or utilized in any form or by any means, electronic or mechanical, including photocopying, recording, or by any information storage and retrieval system, without permission in writing from the publishers.
Printed in Great Britain

# PREFACE

This volume presents the proceedings of the Conference on Hyperfine Interactions Detected by Nuclear Radiation held at Rehovot and Jerusalem in September, 1970. The meeting was conceived as a sequel and continuation of the Asilomar conference in 1967 on the same subject, and the choice of the subject matter was to a large extent guided by that conference.

Several significant new subjects in the field of hyperfine interactions which were first introduced either at the Asilomar conference or shortly afterwards were thoroughly and extensively covered in the present meeting. These subjects are: transient fields acting on ions moving inside magnetized ferromagnetic materials, hyperfine interactions in highly ionized isolated atoms, NMR studies on short-lived nuclear states, precession measurements of nuclear magnetic moments by means of spin rotation and stroboscopy, and the concept of local magnetic moments. In several instances, this was the first occasion for the novel ideas to be developed in a comprehensive way and the subject as a whole to be presented, discussed and brought up to date. Considerable progress was also reported in other longer established subjects. This refers, in particular, to implantation studies, internal magnetic fields at impurities in ferromagnetic materials, channeling, static nuclear orientation and orientation by optical pumping, perturbed angular correlation and the various aspects of Mössbauer studies of hyperfine structure. Included in the program were also a number of general surveys of subjects closely related to the subject matter of the conference: radiation damage from implanted ions, mesic atoms, a critique of quadrupole moment measurements by Coulomb excitation and a survey of nuclear magnetic

## PREFACE

moments in shell model interpretation.

With these up-to-date presentations of subjects of current interest and the comprehensive surveys we hope that this book will serve as a useful reference and as a basic introductory book to newcomers to the field. With this purpose in mind we have also followed the example of the proceedings of the Asilomar conference and included in the book up-to-date tables of nuclear moments, hyperfine fields and isomer shifts.

The presentation of the material in this book closely follows the procedure of the conference in order and organization. The invited papers are followed by the orally presented contributed papers. All other contributed papers are brought under the appropriate session. The contributed papers were originally submitted in a brief version for distribution at the conference. Those papers that were not subsequently extended by the authors are presented in the book in this brief form. The discussions that followed the papers presented at the conference are reproduced, although somewhat abbreviated for editorial reasons.

The conference was organized, directed and conducted by the organizing committee. Although only a small number of the members were involved in the day-to-day work of the conference, all were consulted on matters of general policy and on a variety of specific problems that came up in the course of the preparations. It is the combined judgement and expert knowledge of the organizing committee that set the guidelines and the standards of the conference.

We acknowledge with gratitude the financial aid and good will of the sponsors of the conference and of many other institutions, among those the Israeli Atomic Energy Commission, which made their transportation facilities available to us, the Israeli Foreign Ministry, which provided us with a grant that enabled physicists from India to participate at the conference, the Israeli Ministry of Tourism and the companies "Elscint" of Haifa,

# PREFACE

"Beta" of Beer-Sheva, "Seforad" of the Jordan Valley and the "High Voltage Engineering Corporation" of Burlington, Mass.

The public relations departments of the Weizmann Institute and the Hebrew University kindly and willingly offered us their expert help and advice in the detailed planning and running of the conference. We are especially indebted to Mr. Yitzhak Berman of the Weizmann Institute and to Mrs. Judith Bernstein, the Conference Secretary, for their sustained and patient handling of a task that was always difficult and very often trying.

G. Goldring<br>
R. Kalish

Rehovot, September 10, 1970

## ORGANIZING COMMITTEE

*Chairman*

G. GOLDRING          Weizmann Institute of Science, Rehovot

*Scientific Secretary*

R. KALISH            Technion, Israel Institute of Technology, Haifa

H. BERNAS            Institut de Physique Nucléaire, Orsay

S.K. BHATTACHERJEE   Tata Institute of Fundamental Research, Bombay

F. BOEHM             California Institute of Technology, Pasadena, Calif.

L. GRODZINS          Massachusetts Institute of Technology, Cambridge, Mass.

S.S. HANNA           Stanford University, Stanford, Calif.

E. KARLSSON          University of Uppsala, Uppsala

P. KIENLE            Technische Hochschule, Munich

E. MATTHIAS          Freie Universität, Berlin

I. NOWIK             Hebrew University, Jerusalem

S.L. RUBY            Argonne National Laboratory, Argonne, Illinois

A. SCHWARZSCHILD     Brookhaven National Laboratory, Upton, N.Y.

D.A. SHIRLEY         Lawrence Radiation Laboratory, Berkeley

G.D. SPROUSE        State University of New York,
                    Stony Brook, N.Y.

R.M. STEFFEN        Purdue University, Lafayette,
                    Indiana

N.J. STONE          Oxford University, Oxford


SPONSORS

The International Union of Pure and Applied Physics
The Weizmann Institute of Science, Rehovot
The Hebrew University of Jerusalem
Technion-Israel Institute of Technology, Haifa
The Israel Academy of Sciences and Humanities

CONTENTS

VOLUME 1

PREFACE

**SESSION 1:** HYPERFINE INTERACTION IN SOLIDS, I
   Chairman: H. Bernas

Radiation damage from implanted ions                                    3
   D. Dautreppe

Implantation into ferromagnetic materials                              31
   R.R. Borchers

Implantation into ferromagnetic materials                              75
and the magnetic saturation
   A. Aharoni

The hyperfine fields at $^{42}$Ca in ferromagnetic                     83
metals and the g-factor of the 3.19 MeV $6^+$ state
   M. Marmor, S. Cochavi and D.B. Fossan

Mössbauer and nuclear orientation studies on                           89
atoms implanted in metals
   H. de Waard, P. Schurer, P. Inia,
   L. Niesen and Y.K. Agarwal

Hyperfine field of $^{19}$F ions implanted into a                      96
cubic cobalt lattice
   O. Klepper and F. Bosch

A perturbed angular correlation study of Ta in                         99
Fe, Ni, Cu and Mg
   J.A. Cameron, P.R. Gardner, L. Keszthelyi,
   W.V. Prestwich, Z. Zámori and D.C. Santry

Investigation of radiation damage in Re compounds    102
by angular correlations
    T. Bădică, S. Dima, A. Gelberg, E. Ianovici,
    R. Ion-Mihai and N.G. Zaitseva

The hyperfine magnetic interaction of the first    104
($2^+$) state of $^{142}$Ce in iron, cobalt and nickel
    H.W. Kugel, T. Polga, R. Kalish and
    R.R. Borchers

The range of $^{150}$Sm ions in copper    107
    G.E. Cohn, T. Polga, H.W. Kugel and
    R.R. Borchers

Application of the $5/2^-$-state in $^{63}$Ni to HFS-    110
measurements
    J. Bleck, R. Michaelsen, W. Ribbe and
    W. Zeitz

Nuclear orientation of $^{175}$Yb in Fe and Au and    113
$^{137}$Ce$^m$ in Fe and Ni using ion implantation
    D. Spanjaard, R.A. Fox, J.D. Marsh and
    N.J. Stone

Magnetization direction at rare earth nuclei    119
implanted into ferromagnetic matrices
    I. Ben-Zvi, P. Gilad, M.B. Goldberg,
    G. Goldring, K.-H. Speidel and R. Kalish

Precession measurements of $6^+$ and $8^+$ rotational    123
states in even Dy isotopes
    R. Kalish and W.J. Kossler

Magnetic moments of excited ground-band-and-gamma-    129
band-states in $^{188,192}$Os
    S.H. Sie, I.A. Fraser, J.S. Greenberg,
    A.H. Shaw, R.G. Stokstad and D.A. Bromley

**SESSION 2:**    HYPERFINE INTERACTION IN SOLIDS, II
    Chairman: A. Sunyar

Location of implanted atoms via channeling    137
    B.I. Deutch

Quadrupole effects in NMR spectra on short-lived   167
β-radioactive nuclei, $^{12}$B and $^{12}$N
    K. Sugimoto, A. Mizobuchi, K. Matuda and
    T. Minamisono

Lattice location of Bi and Tl implanted into iron   174
single crystals
    L.C. Feldman, W.M. Augustyniak and
    E.N. Kaufmann

The interpretation of channelling experiments for   179
implanted atom location in crystals
    R.B. Alexander, G. Dearnaley, D.V. Morgan
    and J.M. Poate

Local magnetic moments   185
    B.B. Schwartz

Influence of the microstructure on hyperfine   216
interactions in dilute Yb-Ag alloys
    R. Béraud, I. Berkes, J. Danière,
    R. Haroutunian, G. Marest and R. Rougny

Temperature dependence of the hyperfine inter-   221
action of $^{169}$Tm in Fe: study of slow relaxation
by IPAC
    H. Bernas and H. Gabriel

Channeling studies on iron single crystals and the   226
lattice location of xenon implanted into iron
    L.C. Feldman and D.E. Murnick

Location of Au, Sb and Ce implanted into Fe   229
single crystals
    R.B. Alexander, N.J. Stone, D.V. Morgan
    and J.M. Poate

**SESSION 3:**   HYPERFINE INTERACTION AND NUCLEAR
ORIENTATION, I
    Chairman: M.A. Grace

A review of recent developments in low tempera-   237
ture nuclear orientation
    N.J. Stone

Spinrotation, stroboscopy and dynamic pertur-     291
bations
    E. Recknagel

γ-radiation NMR in beam on the 20.2 msec level     313
in $^{71}$Ge and the 159 μsec level in $^{115}$Sn
    D. Riegel, N. Bräuer, B. Focke and
    E. Matthias

Relaxation phenomena and nuclear magnetic reson-     317
ance of the β-emitters $^{116}$In (14s) and $^{110}$Ag (24s)
produced by capture of polarized neutrons
    H. Ackermann, D. Dubbers, J. Mertens,
    A. Winnacker and P. von Blanckenhagen

Magnetic moment of β-emitter $^{29}$P     325
    K. Sugimoto, A. Mizobuchi and T. Minamisono

Stroboscopic determination of the g factor of     332
the $8^+$ state of $^{208}$Po
    S. Nagamiya, T. Nomura and T. Yamazaki

Nuclear orientation of $^{82}$Br in iron     335
    A.T. Hirshfeld, D.D. Hoppes, W.B. Mann
    and F.J. Schima

NMR/ON of $^{110}$Ag$^m$ in Fe and Ni and $^{96}$Tc in Fe     339
    R.A. Fox, P.D. Johnston, C.J. Sanctuary
    and N.J. Stone

Nuclear spin lattice relaxation measurements     345
below 0.1°K by nuclear orientation
    D. Spanjaard, R.A. Fox, I.R. Williams and
    N.J. Stone

A study of nuclear relaxation at low temperatures     351
in the metastable state of $^{109}$Ag
    N.J. Stone, R.A. Fox, F. Hartmann-Boutron
    and D. Spanjaard

Measurement of the g factor of the 469 keV 11/2⁻     356
level in $^{109}$Cd by stroboscopic observation of
the Larmor precession
    D.H. Bloch, D. Frosch, E.J. Jaeschke,
    H. Pauli and E. Rinsdorf

Determination of the internal magnetic field in     358
pure vanadium with nuclear orientation and
polarized neutrons
    H. Postma, L. Vanneste and V.L. Sailor

# VOLUME 2

**SESSION 4:** HYPERFINE INTERACTION AND NUCLEAR ORIENTATION, II
Chairman: C. Broude

Dynamic polarization by optical pumping    363
    E.W. Otten

Hyperfine field studies in ferromagnetic lattices by PAC using radioactivity    389
    S.K. Bhattacherjee

Excited state NMR measurements on the 14.4 keV state of $^{57}$Fe    433
    N.D. Heiman and J.C. Walker

The influence of defects and of impurities on the E.F.G. parameters in metallic hafnium    439
    J. Berthier, P. Boyer and J.I. Vargas

Hyperfine interactions in hafnium single crystals observed by time differential angular correlation measurements    449
    R.M. Lieder, N. Buttler, K. Killig, K. Beck and E. Bodenstedt

Perturbed angular correlation experiments on $^{100}$Rh in a Ni host near the Curie point    457
    R.C. Reno and C. Hohenemser

A new method for differential perturbed angular correlation measurements    462
    R.S. Raghavan, P. Raghavan and P. Sperr

Temperature dependence of electric-quadrupole  464
hyperfine interactions at ytterbium impurities
in thulium metal
    A.R. Chuhran, A. Li-Scholz and R.L. Rasera

Hyperfine field for Pd, Cd and Te in gadolinium  466
    L. Boström, K. Johansson, E. Karlsson and
    L.-O. Norlin

Theory for TDPAC in non-coaxial electric and  468
magnetic fields
    L. Boström, E. Karlsson and S. Zetterlund

Effects of co-axial electric and magnetic fields  471
produced by polarizing the 4f-shell on time-
integral PAC at low temperatures
    E. Karlsson, M.M. Bajaj and L. Boström

Temperature dependence of the magnetic hyperfine  475
field for Pd in Fe, Ni and Co and for Ru in Fe
    L.-O. Norlin, K. Johansson, E. Karlsson
    and M.R. Ahmed

Investigation of statistically perturbed angular  479
correlations of $^{160}$Dy and $^{172}$Yb at different
temperatures
    H.F. Wagner and M. Forker

Investigation of statistically perturbed angular  487
correlations of $^{154}$Gd and $^{156}$Gd
    H.F. Wagner, M. Forker and U. Weigand

Determination of nuclear spin relaxation times  491
by statistically perturbed γγ-angular correlations
    M. Popp and H.F. Wagner

Hyperfine field of Sc in a Ni host  495
    H.C. Benski and C. Hohenemser

Orientation of the residual nucleus of the reaction  501
$^{56}$Fe(d,p)$^{57}$Fe*(14 keV)$^{57}$Fe
    H. Appel, G. Bueche and W. Renz

Knight shift measurements in 2% Rh-Pd alloys     507
   G.N. Rao

Hyperfine structure and isotope shift of radio-     511
active $^{193}$Hg detected by optical pumping technique
   G. Fulop, C.H. Liu and H.H. Stroke

A fifty-six channel goniometer     513
   T.R. Gerholm, B.G. Pettersson, C. Bargholtz,
   Z.H. Cho, L. Eriksson and L. Gidefeldt

Variation of perturbed angular distribution with     518
the bandwidth of the exciting spectrum
   U. Isaak and G.R. Isaak

## SESSION 5: ISOMER SHIFT AND MESIC ATOMS
   Chairman: S. Ofer

Evaluation of isomer shifts     523
   G.M. Kalvius

Systematics of isomer shifts in some 4d and 5d     595
elements
   G. Kaindl, D. Kucheida, W. Potzel,
   F.E. Wagner, U. Zahn and R.L. Mössbauer

Mössbauer isomer shifts and changes of nuclear     603
charge radii in $^{182,184,186}$W
   F.E. Wagner, H. Schaller, R. Felscher,
   G. Kaindl and P. Kienle

Nuclear charge radius change in $^{161}$Dy     613
   R.L. Cohen and K.W. West

Mesic atoms     619
   S. Devons

Hyperfine anomalies     651
   G.J. Perlow

| | |
|---|---|
| Information on the origin of hyperfine fields in iridium from the hyperfine anomaly of $^{193}$Ir<br>F.E. Wagner and W. Potzel | 681 |
| Isomer shifts and change of the nuclear charge radii of $^{99}$Ru and $^{101}$Ru<br>W. Potzel, F.E. Wagner, G. Kaindl, R.L. Mössbauer and E. Seltzer | 691 |
| Change of nuclear charge radius in even and odd Yb isotopes<br>G.K. Shenoy, G.M. Kalvius, W. Henning, G. Baehre and P. Kienle | 699 |
| Interpretation of the $^{57}$Fe isomer shift by means of atomic Hartree-Fock calculations on a number of ionic states<br>J. Blomqvist, B. Roos and M. Sundbom | 706 |
| Electron densities and isomer shifts for various neptunium ions<br>B.D. Dunlap, G.K. Shenoy, G.M. Kalvius, D. Cohen and J.B. Mann | 709 |
| Isomer shifts of $^{73}$Ge in Fe-Ge and Mn-Ge alloys<br>D. Seyboth, H. Kilian and H. Jena | 718 |
| Solid state and nuclear results in $^{149}$Sm Mössbauer measurements<br>M. Eibschütz, R.L. Cohen and J.H. Wernick | 720 |
| Isomer shifts in muonic atoms<br>H. Backe, R. Engfer, U. Jahnke, E. Kankeleit, K.H. Lindenberger, C. Petitjean, H. Schneuwly, W.U. Schröder, H.K. Walter and K. Wien | 723 |
| Observation of E2 X-ray transitions in muonic atoms<br>H. Backe, R. Engfer, U. Jahnke, E. Kankeleit, K.H. Lindenberger, C. Petitjean, H. Schneuwly, W.U. Schröder and H.K. Walter | 729 |

# VOLUME 3

**SESSION 6:** MÖSSBAUER EFFECT IN TRANSITION ELEMENTS
Chairman: P. Kienle

| | |
|---|---|
| Relaxation effects in Mössbauer spectra<br>H.H. Wickman | 735 |
| Relaxation phenomena detected by Mössbauer effect measurements in $FeNb_2S_4$<br>E. Hermon, H. Rabbie and S. Shtrikman | 763 |
| Hyperfine structure in f shell elements<br>I. Nowik | 769 |
| Magnetic hyperfine fields on holmium and lutetium in Ho-Lu alloys<br>W. Schott and V.L. Sailor | 787 |
| On the magnetic field and charge density of conduction electrons at nuclei in rare earth metals<br>W. Henning, G. Bähre and P. Kienle | 795 |
| Hyperfine interaction in d-shell elements<br>C.E. Johnson | 803 |
| Mössbauer effect in mixed Er iron garnets at 4.2°K<br>U. Atzmony, F.T. Parker, J.C. Walker and E.L. Loh | 816 |
| Mössbauer effect measurements on DyP, DyAs, DySb<br>G. Vécsey and W. Dey | 818 |

Interference between hyperfine-structure states 821
in Mössbauer scattering
    K. Gabathuler and H.J. Leisi

Mössbauer effect studies of relaxation phenomena 827
in $NH_4Fe(SO_4)_2 \cdot 12H_2O$
    S. Mørup and N. Thrane

The internal magnetic field and covalency in $KNiF_3$ 832
and the nickel dihalides
    J.C. Travis and J.J. Spijkerman

Mössbauer investigation of Jahn-Teller distortion 834
spinelic manganites
    G. Filoti, A. Gelberg, V. Gomolea and
    M. Rosenberg

Crystal field effects for $Er^{3+}$ in $HoAl_2$ studied 836
by PAC
    R. Wäppling, E. Karlsson, G. Carlsson and
    M.M. Bajaj

Isomer shifts and hyperfine splitting of the 839
145 keV Mössbauer line of $^{141}Pr$
    W. Kapfhammer, W. Maurer, F.E. Wagner
    and P. Kienle

Effective magnetic fields and isomer shifts at 846
$^{61}Ni$ nuclei in Ni-Pd alloys
    F.E. Obenshain, W. Gläser, G. Czjzek and
    J.E. Tansil

Average hyperfine fields at $^{106}Pd$ nuclei in Ni-Pd 849
alloys
    M.M. El-Shishini, R.W. Lide, P.G. Huray
    and J.O. Thomson

Study of hyperfine interactions in neodymium 853
compounds using the Mössbauer effect
    F.T. Parker and J.C. Walker

Electric quadrupole interaction of $^{178}$Hf in    861
various complex fluorine compounds
   - E. Gerdau, B. Scharnberg and H. Winkler

Angular distribution, line position and line    869
width of 14.4 keV radiation in Fe-Pd alloys
   G.R. Isaak and U. Isaak

Studies of Mössbauer fractions and HFS in frozen    873
solutions
   A. Simopoulos, H.H. Wickman, D. Petridis
   and A. Kostikas

Hyperfine interactions in $Yb_6Fe_{23}$    878
   G. Goretzki, G. Crecelius and S. Hüfner

Mössbauer spectroscopic investigation of the    881
reversible oxygenation of hemoglobin
   A. Trautwein and P. Schretzmann

Temperature dependence of hyperfine fields in    888
$DyCo_2$, $DyNi_2$ and Dy-metal
   - E. Loh, F.T. Parker and J.C. Walker

## SESSION 7: HYPERFINE INTERACTION IN STRIPPED ATOMS, I
   Chairman: S.S. Hanna

Recoil into vacuum    893
   R. Nordhagen

Hyperfine interactions in trans-lead elements    921
via $\alpha$-$\gamma$ PAC
   E.J. Ansaldo and L. Grodzins

Hyperfine interactions of fast recoil nuclei in    931
gas
   G.D. Sprouse

Time dependent angular correlation measurements    961
for the first $2^+$ state of $^{150}$Sm recoiling into
vacuum
   T. Polga, W.M. Roney, H.W. Kugel and
   R.R. Borchers

Velocity dependence of the attenuation mechanism    964
for highly ionized ions recoiling into vacuum
    J. de Boer, J.D. Rogers and S. Steadman

Time differential perturbed γ-ray angular dis-    966
tribution from Yb nuclei recoiling into vacuum
    R. Brenn, L. Lehmann and H. Spehl

Deorientation measurement in $^{20}$Ne    968
    M.M. Faessler, B. Povh and D. Schwalm

Determination of cross section for charge    970
exchange and depolarization in gases from
perturbed angular correlations
    F.N. Gygax and H.J. Leisi

Possible TDPAC following low-energy recoil into    973
vacuum
    R. Armbruster, Y. Dar, J. Gerber and
    J.P. Vivien

Magnetic moment determination at pre-set travel    977
times in the recoil-into-gas method
    T.R. Miller, G.D. Sprouse, P.D. Bond,
    M. Takeda, W.A. Little and S.S. Hanna

Magnetic moment determination by line shape    981
analysis in the recoil-into-gas method
    P.D. Bond, G.D. Sprouse, T.R. Miller and
    S.S. Hanna

**SESSION 8:** HYPERFINE INTERACTION IN STRIPPED ATOMS, II
Chairman: L. Grodzins

Recoil into nonmagnetic metals     987
B. Herskind

Implantation perturbed angular correlation studies     1033
of $^{182,184,186}$W in gadolinium single crystals
E.N. Kaufmann, D.E. Murnick, C.T. Alonso
and L. Grodzins

Quadrupole moment ratio of the first excited     1037
states in $^{192}$Os and $^{190}$Os
R. Avida, I. Ben-Zvi, P. Gilad, M. Goldberg,
G. Goldring, K.-H. Speidel, A. Sprinzak and
Z. Vager

Measurements of time dependent perturbed angular     1043
correlations of rare earth nuclei implanted in
copper
P. Ryge, H.W. Kugel and R.R. Borchers

Transient fields     1055
A. Winther

Angular distributions perturbed by quadrupole     1065
interactions in axially symmetric single crystals
L. Grodzins, C. Alonso and J. Alonso

The measurement of the magnitude and sign of the  1067
nuclear quadrupole interaction via the spin precession of an angular correlation
    L. Grodzins and O. Klepper

Time differential studies of $^{166}$Er in copper  1077
    G.M. Heestand, K. Bonde Nielsen, H. Ravn,
    F. Abildskov and B.I. Deutch

Transient fields on $^{103}$Rh in Fe at low recoil  1081
energies
    S.K. Bhattacherjee, H.G. Devare, H.C. Jain,
    M.C. Joshi and C.V.K. Baba

## SESSION 9: NUCLEAR PHYSICS
    Chairman: S.G. Cohen

The quadrupole moment of $^{114}$Cd: a test case for  1091
the reorientation effect
    U. Smilansky

Summary of nuclear moments reported at the  1119
conference
    R. Kalish

Nuclear magnetic moments - shell model interpret-  1133
ation
    I. Talmi

Magnetic moment of the 2761-keV $8^+$ state in $^{92}$Mo  1170
    S. Cochavi, J.M. McDonald and D.B. Fossan

Halflife and magnetic moment of 93 keV $(9/2^+)$  1173
level in $^{103}$Rh
    C.V.K. Baba, S.K. Bhattacherjee and H.C. Jain

Anomalous orbital magnetism of proton deduced from  1177
the g-factor of the $11^-$ state of $^{210}$Po
    T. Yamazaki, T. Nomura, S. Nagamiya and
    T. Katou

g-factor measurements for the 160 μs state in $^{115}$Sn and 340 μs state in $^{117}$Sb excited by pulsed beam bombardments on liquid metallic targets ... 1180
    E.A. Ivanov and G. Pascovici

IMPAC measurements on levels of $^{125}$Te ... 1182
    W.M. Roney and R.R. Borchers

The magnetic moments of the 7/2⁻ mirror states in $^{37}$Ar and $^{37}$K ... 1185
    W.L. Randolph, Jr., and R.R. Borchers

Magnetic moment of the 10 μsec level in $^{58}$Co and relaxation of aligned $^{58}$Co nuclei implanted in α- and γ-iron ... 1192
    M. Becker, H. Bertschat, H.-E. Mahnke, E. Recknagel, R. Sielemann, B. Spellmeyer and Th. Wickert

Magnetic hyperfine interaction measurements on the first excited state in $^{62}$Cu ... 1195
    J. Bleck, R. Michaelsen, W. Ribbe and W.-D. Zeitz

## TABLES

Table of change in nuclear mean square charge radius ... 1201
    G.K. Shenoy and G.M. Kalvius

Table of hyperfine fields ... 1239
    T.A. Koster and D.A. Shirley

Table of nuclear moments ... 1255
    V.S. Shirley

## AUTHOR INDEX ... 1337

## LIST OF PARTICIPANTS ... 1341

SESSION 6

MÖSSBAUER EFFECT

IN

TRANSITION ELEMENTS

Chairman:

P. KIENLE

*Technischen Hochschule München*

# RELAXATION EFFECTS IN MÖSSBAUER SPECTRA[*]

H.H. WICKMAN[†]

*Nuclear Research Center "Democritos,"*

*Athens, Greece*

ABSTRACT

Several recent experimental and theoretical studies of solid state relaxation phenomena are reviewed. Emphasis is placed on (1) magnetic relaxation as a prototype electronic process affecting internal fields, and (2) solid state glass transformation during which the recoil free fraction may exhibit time dependent features.

I. INTRODUCTION

It is well known that from the nuclear standpoint, all parameters associated with a Mössbauer absorption pattern may be perturbed by suitable time-dependent processes in the atomic or lattice environment of the absorbing system. Thus, the recoil free fraction, or any

---

[*]Invited paper.

[†]Permanent address: Bell Telephone Labs., Murray Hill, New Jersey, U.S.A.

of the electron-nucleus multipole interactions may reflect dynamic inter- or intra-molecular events. (One might also include relativistic effects, such as gravity fluctuations, as sources of line broadening, but such non solid-state factors will not be considered here.) In the following, we note a few examples where electronic relaxation processes are associated with changes in the shape or intensity of Mössbauer spectra. Since most recent studies have dealt with magnetic relaxation phenomena, this type of interaction will receive the major emphasis.

The origins of strictly electronic relaxation effects in Mössbauer spectra are of course no different from the time-dependent processes affecting angular correlations, nuclear- and electron-magnetic-resonance, etc. Extensive reviews of the basic notions are readily available [1-6]. In most cases the stochastic nature of the problem appears as a characteristic time $\tau_c$, or relaxation rate $\Omega \equiv \tau_c^{-1}$, for a dynamical variable. In addition, a specific case will be described by time-independent factors expressing the strength of the multipole interactions [7]. For quadrupole and magnetic dipole interactions, the static field strength may be assigned the frequencies $\Omega_Q$ and $\Omega_H$. Finally, the natural linewidth of the oscillator is important and is related to the mean nuclear-lifetime, $\Gamma \equiv \tau_N^{-1}$. The latter width is often large compared with nucleus-nucleus magnetic dipole interactions, so field fluctuations from nearby nuclear moments are not easily observed. On the other hand, when fluctuation rates in the strong electronic sources of the multipole interactions approach the characteristic frequencies $\Omega_Q$ or $\Omega_H$, the latter lose precise meaning spectroscopically, the hyperfine lines initially broaden, and the whole hyperfine pattern may eventually collapse when the time average of the field source vanishes over the nuclear lifetime. Examples of such behavior are of course well known in paramagnetic relaxation spectra where magnetic couplings are affected. A case where quadrupole couplings are perturbed occurs when random strains cause fluctuating electric field gradients; such effects have been observed in $Fe^{2+}$ doped MgO [8-10].

It may also happen that the monopole interaction fluctuates owing to limited lifetimes of electronic charge states, or due to thermal fluctuations among electronic levels producing different net charge densities at the nucleus. However, well defined examples of such behavior are in practice difficult to establish. Finally, the recoil free fraction itself may have a time dependent character, depending on local heating and diffusion effects [11, 12], or in some circumstances resulting from phase transformations [13-16]. In the latter cases, the characteristic times involved may not have the simple significance mentioned earlier.

The next Section introduces some theoretical considerations, and the following two Sections deal with experimental studies of time-dependent phenomena. The first deals with magnetic relaxation spectra, and the second with bulk properties related to transformations occurring in glass forming materials.

## II. THEORY

The hyperfine structure for a particular electronic-nuclear level system may be summarized by a spin Hamiltonian of the form [17]

$$H = H_M(\vec{S}, \vec{L}, \vec{I}) + H_Q(\vec{L}, \vec{I}) + H'(\vec{S}, \vec{L}, \vec{R}) \qquad (1)$$

The notation is meant simply to indicate that (i) the magnetic hfs arises from a coupling of the nuclear moment to the spin and orbital character of the electrons, and (ii) the ionic source of the efg arises from the orbital angular momentum of the electronic system (we neglect at this time the lattice contribution to the efg). The third term represents the coupling between the electrons and lattice or spin reservoirs. This coupling leads to spin-lattice or spin-spin relaxation, and hence to an overall time dependence to the motion of the local electron-nucleus spin system. As noted earlier, the nuclear spin-spin and spin-lattice relaxation processes are generally much weaker, though probably not entirely negligible in all cases. In the following, however, this relaxation channel will be neglected.

The time dependent features of Eq. (1), and their effect on Mössbauer spectra, were first studied by Afanas'ev and Kagan [1]. Additional approaches to the problem have since appeared and have been reviewed elsewhere [3-6]. The methods include perturbation treatments, rate equations, and stochastic models. While early discussions mainly dealt with situations where $I_z$ was a constant of motion, i.e., where $[I_z, H] = 0$, more recent analyses have generalized the models to cover situations where the Hamiltonians $H_M + H_Q$ have different eigenfunctions in the limits of fast and slow electronic relaxation times. For example, Hirst [18] has described a generalization of the rate equation method and has used this method to analyze relaxation spectra observed in $^{166}$Er in zirconium metal [19]. A very general method, based on Liouville operator techniques and stationary Markov stochastic models, has been described by Gabriel, Bosse and Rander [6, 20], and Blume [21]. From the standpoint of generality and elegance, this technique is most attractive. The basic features of the method, starting from a correlation function, are probably most easily understood by reference to the discussion of Blume [21]. The formal aspects of the problem are given in some detail by Gabriel [6] and Bosse [22].

We summarize here a few features of the discussion given by Blume, and later give examples predicted by this formulation (or equivalently by the methods of Gabriel et al. [6] or Blume and Tjon [23]). When it is possible to describe the interaction in Eq. (1) by classical fields (of electronic origin) which may fluctuate among a finite number of principal values and orientations, the Hamiltonian may be rewritten in the form

$$H(t) = \sum_j \hat{V}_j f_j(t) \qquad (2)$$

The $\hat{V}_j$ represent nuclear quantum mechanical operators corresponding to dipole and quadrupole interactions. The $F_j$ are functions of a random variable $f(t)$ which posseses a finite range corresponding to the states $s_1, s_2, \ldots, s_n$. Each state corresponds to a particular field configuration and an important part of the problem is to specify initial probabilities for these states, and the transition probabilities among them. Equation (2) is a formal way of saying that if we restrict our attention to the nucleus, the effect of relaxation in Eq. (1) may be reviewed as random changes in the magnitude and direction of the interactions from the extra-nuclear electrons.

The stochastic nature of the problem arises explicitly in the evaluation of the line shapes of the Mössbauer pattern. The latter is expressed as the Fourier transform of the correlation function of the nuclear radiation operator $H^{(+)}$:

$$I(\omega) = (2/\Gamma) \, \text{Re} \int_0^\infty dt \, \exp(i\omega t - \tfrac{1}{2}\Gamma t)(<H^{(-)}H^{(+)}(t)>)_{av} \quad (3)$$

The bra-kets in the integrand denote occupational averages over eigenstates of the Hamiltonian in Eq. (2). The parentheses express a stochastic average which must be performed when fluctuations among the "long-relaxation limit" eigenstates become important. The nuclear radiation operator is expressed in the Heisenberg representation [20-24]

$$H^{(+)}(t) = e^{i\int_0^t H(t')dt'} H^{(+)} e^{-i\int_0^t H(t')dt'} \quad (4)$$

At this point, a Liouville operator $H^x$ corresponding to $H$ is introduced. This can be used to effectively unscramble the integrations appearing in Eq. (4) [25]. Using a restricted averaging process [26], the integrand of Eq. (3) may then be evaluated. These techniques are not required when $[I_z, H(t)] = 0$, for all t, but are essentially indispensable where this condition is not satisfied. Since many Mössbauer situations fall in the latter category, the results of the theories are very useful. No attempt is made here to describe the methods further. However, it is instructive to display the final form of Eq. (3), when the operations have been carried out. In the form given by Blume [21] it is

$$I(\omega) \propto (2I_1+1) \text{ Re} \sum_{\substack{m_1,m_1' \\ m_0,m_0'}} \langle I_1 m_1 | H^{(-)} | I_0 m_0 \rangle \langle I_0 m_0' | H^{(+)} | I_1 m_1' \rangle$$

$$* \sum_{a,b} P_a \langle I_0 m_0 I_1 m_1 a | [(-i\omega + \tfrac{1}{2}\Gamma)\tilde{1} - \tilde{W} -$$

$$- i \sum_j \tilde{V}_j^x \tilde{F}_j ]^{-1} | I_0 m_0' I_1 m_1 b \rangle. \qquad (5)$$

The final operation is the inversion of a matrix sum labeled by quantum mechanical ($m_1$, $m_0$) and stochastic (a, b) variables. Appearing in the sum are a trivial frequency and linewidth term ($-i\omega + \tfrac{1}{2}\Gamma$), the matrix $\tilde{W}$ expressing the transition probabilities among the Markov states introduced above, and finally the sum of Liouville operators corresponding the stochastic Hamiltonian ($H \to H^x$) of Eq. (2). In all cases of interest, a large number of matrix elements vanish, since the quantum mechanical operators are diagonal with respect to stochastic operators (appearing as the matrices $F$), and vice versa. The significance of the various matrices of Eq. (5) is most easily illustrated by references to specific cases, one of which is found in [21], and another given below.

## III. EXAMPLES

### A. *Magnetic Relaxation*

There are a few types of paramagnetic hyperfine interactions and relaxation behavior that account for most of the observed "relaxation spectra". First, of course, are strongly anisotropic Kramers doublets for which the interaction is simply $A_z T_z S_z$ [27-29]. The rare-earth region has many examples [7]. Essentially, all well resolved zero-field paramagnetic hfs (rare-earth or transition metal isotope) is of this type. For other non-strongly anisotropic interactions, local fields generally perturb the static hfs (e.g. that predicted by Eq. 6) and some form of intermediate line shapes are found. If ligand hfs is also present, the net patterns can be quite complicated [30]. Another situation occurs when local fields, and relaxation processes, are greater in strength than nuclear linewidths or hfs energy spacings, but at the same time are small compared to typical electronic Zeeman energies. In these cases an external magnetic field (H = 30 to 500 Oe) produces an electronic polarization and provides a reasonably well defined nuclear quantization axis [31]. The resulting hfs may then be analyzed by an effective field interaction and, usually, intermediate relaxation rates. More generally, magnetic polarization of a fast relaxing paramagnetic sample will trivially lead to a Zeeman pattern whose overall width is proportional to the ionic moment. Relaxation in magnetically ordered systems has only been rarely observed (see below), generally because relaxation rates are either too slow (T → 0°K) or too fast to affect the linewidths of the hfs level [32].

Of the various types of relaxation processes, spin-spin, spin-lattice, and cross-relaxation, the most commonly encountered is electronic spin-spin relaxation. Most cases of natural dilution, occuring when a paramagnetic ion is incorporated in a large molecule, fall in this category. Further dilution (in simple salts to $\sim$ 1 atomic percent) reduces spin-spin couplings and spin-lattice relaxation becomes more important. This usually perturbs the $Fe^{3+}$ hfs in the range 50 to 70°K; below this a well resolved pattern is found. A useful discussion of spin-lattice relaxation in the ferric ion has been given by Suzdalev et al. [33]. In the rare-earth region, where the $A_z I_z S_z$ interaction makes spin-spin processes ineffective even in concentrated salts, the Orbach type spin-lattice relaxation may be studied by the Mössbauer technique [34]. Cross relaxation processes have received little attention to date.

We now give results for two systems where the Mössbauer technique is particularly appropriate for the study of relaxation processes. The first case is divalent $^{151}$Eu at low concentration in a well defined lattice (e.g. CaS) [35]). The second is an example of relaxation in the temperature range where a para- to ferro-magnetic transition occurs in an iron complex [36].

Divalent europium has as ground state an $^8S_{7/2}$ term. It is generally only weakly perturbed by crystal field interactions, and hence is an almost ideal S-state ion. Relaxation processes in ionic solids at low temperatures and with small $Eu^{2+}$ concentration are limited to the coupling (exchange or dipole-dipole) between the spin moments. At concentrations of 1 atomic percent, the couplings are large compared to hyperfine energies, and hence classical EPR studies of relaxation processes of this type require much lower concentrations (<0.5 at .%). The Mössbauer method is not so severely restricted and may be useful in the range 1 to 10 at.%. It thus offers a complementary technique to EPR, and further may be used at lower fields than those employed for magnetic resonance.

As a representative system we take $Eu^{2+}$ doped CaS, the concentration range of $Eu^{2+}$ being from 1 to 4 at.%. The isotropic nature of the intra- and interionic couplings leads to a broadened and structureless $^{151}Eu$ resonance line in zero applied fields. It is therefore convenient to apply a small polarizing field to achieve a quantization axis for the nuclei. The fields required depend on concentration and temperature, but usually lie in the range $0.5 \leq H \leq 2$ kOe. The data analysis is quite direct if one assumes a large polarizing field. This results in a level system of the Paschen-Back type, the electronic Zeeman energies being widely spaced compared with hyperfine or electronic dipole interactions. In this case the fluctuating Hamiltonian consists of simple classical fields $H(M_s)$, whose magnitude and direction depend on the instantaneous electronic quantum state $M_s$. The time dependence enters because of fluctuations between the $M_s$ states, so the Hamiltonian in simplified form becomes

$$H(t) = g\beta H I_z M_s(t). \qquad (6)$$

The transition rates among the 8 electronic levels (2S+1) are given by the spin-spin interaction, and the "Golden Rule."

$$W(M_s \rightarrow M_s') = C |<M_s'|S_\pm|M_s>|^2 \ \bar{n}(M_s') \qquad (7)$$

C is an empirically varied parameter, but in special cases should be computable on *a priori* grounds. The factor $\bar{n}(M_s)$ is the normalized electronic Boltzmann distribution, which implies population effects arising from varying temperature and fields.

The computation of spectra consists of setting up stochastic matrices based on Eq. (7), with the quantum mechanical portion given by Eq. (6). The method of Blume may be readily applied, or in this case the simpler rate equation method is adequate [37]. In either case the results yield equivalent spectra which have shapes depending on temperature, relaxation rate, and external magnetic field. All hyperfine parameters were assumed to be the same as in concentrated europium sulfide. It should be emphasized that the procedure here is entirely analogous to methods employed earlier to discuss relaxation phenomena of $Fe^{3+}$ ions [38, 39]. A difference in complication occurs because $^{151}$Eu has nuclear spins of 7/2 and 5/2. There are thus 18 nuclear transitions (M1 radiation) whose energies are being modulated by electronic transitions.

A comparison of theory and experiment is given in Fig. 1. Concentration, x, is indicated in the Figure, together with the temperature and field strength. Good agreement between theory and data is found when, as assumed, the electronic Zeeman energy is large (H ≥ 5 kOe). An error of ±15% is estimated in the relaxation rate. In lower fields, the effective rate increases markedly. This is due to cross-relaxation effects occuring when the Zeeman energies become more nearly commensurate with the dipolar widths. Because of isotropy in the sample, an approximate average in zero field leads to a line shape consistent with a motionally narrowed pattern; the data were well fit in zero field by a relaxation rate of 150 Mc/sec. For moderate fields (0 < H ≤ 3-5 kOe) the spectra are more complex and we were unable to fit the data with a line shape corresponding to a unique relaxation parameter. This indicates a local distribution of rate processes, the average being about 50 Mc/sec (H = 4.3 kOe, T = 4.18°K).

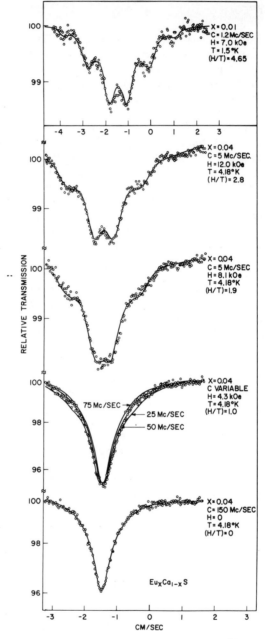

Fig. 1. Mössbauer effect in $Eu^{2+}$:CaS.

RELAXATION EFFECTS                                                    745

The preceding analysis was basically empirical and meant simply to test the ease of applying the Mössbauer technique to the study of Eu paramagnetic relaxation. A logical and much more interesting next step would be to explicitly compute dipole interactions for the statistical distribution of europium ions in the cubic solid. This would yield a dipole contribution to the transition rate, and any difference between theory and experiment would be related to exchange interactions between the $Eu^{2+}$ ions. This coupling in the case of europium chalcogenides is of some interest. Finally, more detailed study and analysis of the magnetic field dependence of the line shapes (using more sophisticated models) would yield the magnetic field dependence of the relaxation rate, together with useful information on europium-europiun cross relaxation. Finally, doped crystals with other rare-earth ion would offer potential for cross-relaxation between Mössbauer and non-Mössbauer ions.

An example of circumstances where the general theories of Gabriel et al. [6], Blume and Tjon [23], or Blume [21] are definitely required is given by the case of a magnetic field which fluctuates about an axis perpendicular to a large electric field gradient. The Hamiltonian of Eq. (2) in this case is given by

$$V_1 = (e^2qQ/4)(3I_x^2 - J(J + 1)) \quad f_1(t) = 1$$
$$V_2 = g_N\beta_N HI_z \quad f_2(t) = f(t) \tag{8}$$

with f(t) randomly assuming the values ±3/2 and ±1/2. A material whose hyperfine interactions are quite accurately described by this Hamiltonian is bis (N, N-diethyl-dithiocarbamato) ferric chloride. This complex is paramagnet to 4.23°K; here it undergoes a ferromagnetic transition. Its Mössbauer spectra show relaxation effects for absorber temperatures between 1.2 and 4.2°K [36]. The quadrupole interaction is large ($e^2qQ/2$ = 0.268 cm/sec) and perpendicular (i.e. principal axis) to the magnetic interaction. A final characteristic of the material is that its electronic spin is 3/2. The electronic levels consist of upper $|\pm1/2\rangle$ and lower $|\pm3/2\rangle$ Kramers doublets. Each is perturbed by an isotropic core polarization (H = 220 kOe in Eq. (6)). The $|\pm 3/2\rangle$ doublet is of the effective field type ($g_z = 6$, $g_x = g_y = 0$) but the $|\pm 1/2\rangle$ level is not ($g_z = 2$, $g_x = g_y = 4$). The model assumes, even in the paramagnetic state, an effective polarization along the z-axis. This assumption is necessary to avoid the complication of explicitly introducing spin quantum variables in the nuclear Hamiltonian. We shall see below that in the present case this assumption does not lead to large errors in the data analysis.

Mössbauer spectra for absorber temperatures in the range 1.2 to 4.2°K are shown in Fig. 2. Also shown are theoretical curves resulting from the model Hamiltonian given above and the stochastic theories discussed earlier. Above $T_c$ the data were fit by assigning the efg interaction and core polarization term (derived from low temperature resolved hfs spectra) together with a spin-spin coupling rate (Eq. (7)). Once these parameters are specified, spectra may be generated for several temperatures; this allows a good determination of the spin-spin coupling constant C in the paramagnetic region. Near and below the critical point, the spin-spin coupling appears to change owing to the onset of spontaneous polarization in the sample. The data were therefore fit with two effective spin-spin coupling constants: One for $T > T_c$ (C = 3kMc) and the other for $T < T_c$ (C = 1 kMc). The intermolecular field was assumed to follow a Brillouin function with $J = 3/2$ and $T_c = 2.43°K$.

That a change in value of the relaxation parameter C should occur on approaching or passing through the magnetic transition point is not entirely unexpected. The origin of this effect is attributed primarily to the onset of polarization in the $|M_s = \pm 1/2\rangle$ excited doublet. Below $T_c$ the four-state model is assumed valid and the derived constant C is taken as a reasonable measure of the electronic spin-spin coupling between iron atoms. At temperatures greater than $T_c$ the effective interaction is roughly one-third its former value. Thus, the analysis indicates that a correction factor of roughly 1/3 should be applied to interaction rates obtained from analysis of paramagnetic hfs in bis (dtc)'s. The correction factor

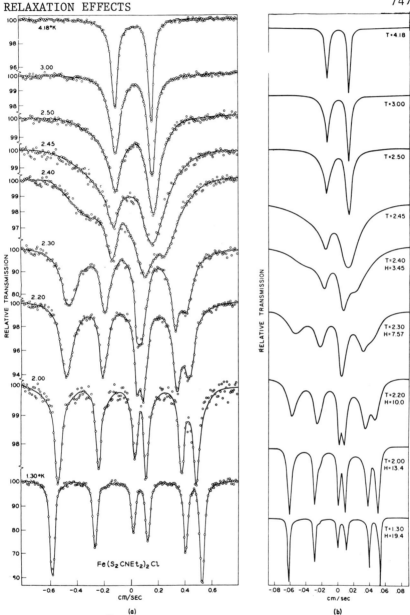

Fig. 2.  (a) Mössbauer effect of polycrystalline Fe(diethyldtc)$_2$Cl,
(b) theory.

applies only to the case where the $|\pm 1/2\rangle$ doublet is highest lying. Here any correction is expected to be minimal, since this level is rapidly depopulated in the temperature range of interest for the Mössbauer relaxation spectra. On the other hand, when the $|\pm 1/2\rangle$ doublet is lowest lying, $S_x(t)$ and $S_y(t)$ behavior may be more important and it is possible that more significant correction factors might be necessary.

B. *"Relaxation Effects" in the Recoil Free Fraction*

When an atom is displaced in a lattice by a fission process, or by an implantation technique, it generally reaches an equilibrium state during an interval in the range 1 to 100 psec [40]. This time is quite short compared to most Mössbauer nuclei lifetimes, so transient processes of this type are not likely to perturb the recoil free fraction (r.f.f.) in an observable fashion. The inverse situation, where the lattice (and only indirectly the impurity) undergoes substantial changes, can represent a condition more likely to result in observable changes. For example, experiments showing modifications in linewidths and probably r.f.f.'s have been found near melting points [12] and in viscous liquids [41]. Time dependent changes of the r.f.f. resulting from well defined crystallographic transformations have, however, not been widely observed.

In this connection, Dezsi et al have recently reported an interesting loss of recoil free fraction of iron ions in frozen aqueous solutions (FAS) [13, 42]. Kaplan and co-workers [14, 43], and Pelah and Ruby [15] have reported additional studies of iron, europium and tin isotopes in FAS. The experimental picture is rather simple. Aqueous solutions of the metal ions are quick frozen, and their Mössbauer pattern then monitored as a function of temperature. Near a characteristic temperature $T_c$, the recoil free fraction either disappears (for periods ranging from minutes to hours) or shows a marked decrease - in both cases in an irreversible fashion. Return to lower temperatures and similar cycling shows no anomalous effects. At $T_c$ the hyperfine pattern or isomer shift may also show a discontinuous variation. This behavior, though not an example of strictly electronic relaxation, nevertheless implies some form of bulk or local relaxation leading to a modulation of the recoil free fraction. Certainly under circumstances of eutectic formation, or *bona fide* melting of the solid, the loss of r.f.f. is not surprising. At the moment, however, most of the reported cases do not seem explicable on this basis. For example, Dezsi et al attributed the loss of r.f.f. to a cubic to hexagonal ice transformation [42]. Nozik and Kaplan came to similar conclusions and proposed a model for the ferrous ion in ice [14]. Recently, Pelah and Ruby have reported Mössbauer and conductivity measurements of tin salts in FAS [15]. Their results were similar to the earlier studies, but their interpretations of the cause of the change of r.f.f. was different. These authors viewed the r.f.f. variation as resulting from a local excitation of the microstructive around the $Sn^{4+}$ ions. This excitation was attributed to mobile charge carriers such as $H^+$ ions. Their motion could be thermally activated from "unstable" low temperature to stable high temperature traps.

Additional information is given by experiments of Di Lorenzo and Kaplan who show perturbations of $T_c$ in FAS containing $Fe^{2+}$ and $Eu^{3+}$ ions [43]. Further, in certain dilute $Fe^{2+}$ solutions, the FAS show a new and lower temperature region, 180°K, where loss of r.f.f. occurs. Above this temperature the effect reappears, only to disappear again at 190°K in the usual fashion [43]. One of the explanations proposed for this behavior was the presence of a vitreous phase which "melts" at 180°K to form the usual cubic ice crystallites.

It is interesting that the above qualitative results, together with the time scales of the events near $T_c$, are rather similar to "relaxation" phenomena observed in viscosity or index of refraction in materials exhibiting glass transition [44]. Although one or two brief references to a low temperature "vitreous" form of ice appear in the above works, no explicit consideration of the implications of a low temperature glass transition in ice, or other materials in which Mössbauer ions are entrapped, seems to have been given. Glassy states of alcohols are easily obtained, and to explore the behavior of Mössbauer parameters at a glass transition, experiments in iron containing methanol glasses have been performed [16].

Before considering the data, it is worth reviewing briefly the physical behavior of a typical material exhibiting a glass transition. The specific heat of glycerol vs. temperature is shown in Fig. 3 [45]. The so-called glass transition is assigned to the abrupt lambda-like decrease in specific heat which occurs on cooling the super-cooled liquid glycerol. On warming, and depending on the particular glass, it is often possible to move reversibly up the liquid-glass curve without crystallization. A second characteristic of a glass is that normally an increase in sample temperature need not immediately lead to a physical state corresponding to the "equilibrium" curve of Fig. 3. In fact, rapid raising or decreasing of the temperature near $T_g$ results in a non-equilibrium state which then (in the absence of crystallization) "relaxes" to the equilibrium condition. The time constant here depends on the particular material but is normally minutes to hours in duration.

Two simple questions may be raised in connection with Mössbauer studies. First, how sensitive is the Mössbauer technique in detection of $T_g$? Second, assuming "quick" Mössbauer experiments can be performed, what is the behavior of the Mössbauer spectrum while the glass solvent is relaxing? Experiments performed thus far have concentrated primarily on the first question, and the present discussion will be restricted to the evidence for $T_g$ in the Mössbauer data. In this connection, it might appear from previous studies in glycerol that the Mössbauer technique is insensitive to $T_g$ [41]. Here there appears to be little, if any, change in the r.f.f. near $T_g$. However, viscosity changes in glycerol are relatively slow at temperatures near $T_g$; in addition, the time scale of the experiments of Champeney and Woodhams [41] was not specified.

A more favorable case is methanol, where the much smaller molecular units imply a greater thermal activity

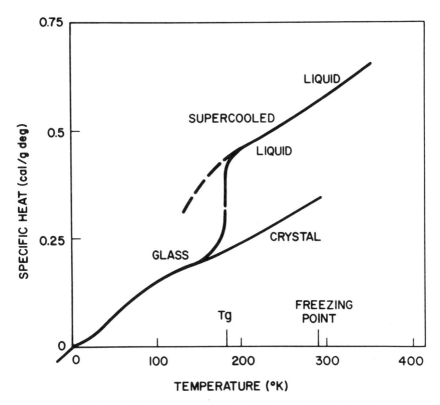

Fig. 3. Temperature variation of the specific heat of glycerol.

near $T_g$. The glass transition in methanol is 102°K [46]. Figure 4 shows the variation of the r.f.f. of the divalent iron in $FeCl_2$ doped methanol. Quick freezing produces clear glasses, and the Mössbauer pattern consists of a quadrupole split doublet with $\Delta E_Q$ = 3.4 mm/sec at 77°K. The quadrupole splitting changes somewhat with temperature, but the change in r.f.f. is more pronounced and more clearly reflects the glass transition of the material. At approximately 104°K, the r.f.f. markedly decreases and goes through a minimum at ∿109°K. The rate

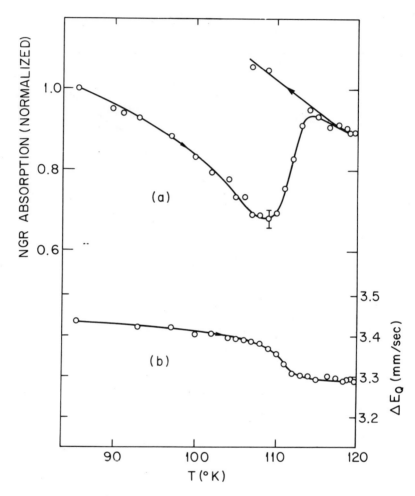

Fig. 4. Temperature variation of the recoil free fraction of FeCl$_2$ doped methanol glass.

of accumulating Mössbauer data was "slow" compared to any relaxation processes in the glass. Thus, the Mössbauer data strongly suggest a pronounced decrease in viscosity near and above $T_g$. For reasons discussed earlier, the assignment of $T_g$ on the basis of the Mössbauer data should be made near the onset of the decrease in r.f.f. Hence, a $T_g$ of $\sim 104°K$ may be derived from the Mössbauer measurements, and this value is in acceptable agreement with the colorimetric result of 102°K [46]. In the absence of crystallization, the r.f.f. would be expected to rapidly fall to zero as the temperature is increased. In methanol, however, it is difficult to maintain the super-cooled liquid and crystallization usually takes place about 2° above $T_g$ [46]. Thus the increase in r.f.f. reflects the crystallization process. The rate of crystallization is evidently not extremely rapid, and the data show a competition between melting of the glass, with concomitant loss of r.f.f. and crystallization which leads to an increase in the r.f.f.

The basic point shown by this data is that a definite coupling of the Mössbauer r.f.f. with the glass transition in methanol has been observed. Extensions of the present work should allow acquisition of more direct information regarding relaxation processes in the non-equilibrium glass states. First, by decreasing the time duration of the experiments, repeated isothermal experiments should be able to detect the effect of the glass relaxation on the microscopic ion environment. A second point is connected with electron relaxation processes in the iron ions near $T_g$. At the dilution and temperature range of the experiments, ferric ions, for example, typically reflect spin-lattice relaxation processes in their paramagnetic hfs. These processes can be quite sensitive to lattice structure variation, and an investigation of $Fe^{3+}$ relaxation spectra in the region of $T_g$ may well be fruitful. Finally, determination of the concentration dependence of the Mössbauer parameters, and of the observed $T_g$, is of interest.

In conclusion, the question of the relation of the present studies in a glass forming material to the previous FAS work may be raised. It is known that ice can form a vitreous state, but this is generally difficult to obtain [47]. Further, epr studies in FAS strongly suggest that during quick freezing, the metal ions segregate in regions which are essentially amorphous [48]. An additional observation is that normal concentrated aqueous solutions of acid or metal salts easily form glassy states upon quick freezing. These facts suggest that the following sequence of events may be important in the FAS. First, an initial glassy state is formed which contains the metal ions. This says nothing about the state of the bulk lattice condition which may, for example, be in one of its normal crystallographic forms. The glass region will melt near its glass temperature, and on the basis of the methanol results, this is sufficient to produce a loss or reduction of the r.f.f. Crystallization, which is irreversible, could then produce a return of the r.f.f. This sequence of events would explain much of the results in FAS.

There are further points, however, which must be considered. First, if a glassy region is surrounded by a homogeneous solid solvent, and the latter undergoes an exo- or endothermic phase transformation, the glass region may indeed appear coupled to the lattice. That is, the glass may simply melt at or near the phase transformation, if the latter is near $T_g$. Second, when a glass region has melted (for any reason) and a recrystallization occurs, one is faced again with the problems of phase separation upon crystallization. A trivial possibility is the formation of a new glassy phase plus crystalline solid. (This of course recalls the original result of quick freezing.) Another point is that in the absence of crystallization, a melted glassy state could result in a rather wide temperature region where no Mössbauer effect is visible.

Finally, we note that a true crystallographic transformation itself is not without the potential of displaying hysteritic effects. These may lead to complications even after all regions of an absorber are crystalline. A good example of such behavior is given by Mössbauer studies of an iron complex reported by Jesson et al [49]. This work shows a ferrous complex which has two crystallographic forms, in one of which the iron is high spin and in the other low spin. However the relative proportions of the two spin states, at a given temperarure, reflect the relative amounts of polymorphs present. The Mössbauer data give rather nice evidence for hysteretic effects associated with the crystallographic transformation. In either case, however, no time-dependent changes in r.f.f. accompany the crystallographic transformation itself.

The above comments are not inclusive, but are intended only to point out typical sources of metastable environments for Mössbauer ions in quick frozen solutions. It appears nearly certain that amorphous regions are of primary importance in FAS studies. However, the potential for variations in detail from system to system cannot be overestimated.

IV. CONCLUSION

A few topics dealing with relaxation processes in solids have been presented. The formalism and discussions connected with electronic relaxation have been shown to be rather complete. Most relaxation spectra may be characterized by a phenomenological relaxation parameter. The latter can then be related quantitatively to specific relaxation processes such as spin-spin, spin-lattice, etc. By contrast, a far less well defined case is given by a study of a simple glass transition in methanol. The process here may be defined only qualitatively. However, the initial experiments suggest a potential for the acquisition of microscopic information directly related to bulk relaxation processes known to occur in the glassy matrix. For this type of investigation, the Mössbauer technique appears uniquely well suited.

REFERENCES

1. AFANAS'EV A.M. and KAGAN, YU., *Zh. Eksper. Teor. Fiz.* 45, 1660 (1963); *Soviet Phys. JETP* 18, 1139 (1964); ibid 47, 1108 (1964); *Soviet Phys. JETP* 20, 743 (1965).

2. VAN ZORGE, B.C., VAN DER WOUDE, F. and CASPERS, W.J., *Z. Physik* 221, 113 (1969).

3. WICKMAN, H.H. and WERTHEIM, G.K., in *Chemical Aplications of the Mössbauer Effect*, V. Goldanskii and

R.H. Herber, Eds. Academic Press (1968).

4.  BLUME, M., in *Hyperfine Structure and Nuclear Radiations*, E. Matthias and D.A. Shirley, Eds. North-Holland Pub. Co. (1968), p. 911.

5.  WEGENER, H. and RITTER, G., *Z. Angew. Physik* 24, 270 (1968).

6.  GABRIEL, H., BOSSE, J. and RANDER, K., *Phys. Stat. Sol.* 27, 301 (1968).

7.  OFER, S., NOWIK, I. and COHEN, S.G., in *Chemical Applications of the Mössbauer Effect*. V. Goldanskii and R.H. Herber, Eds, Academic Press (1968).

8.  LEIDER, H.R. and PIPKORN, D.N., *Phys. Rev.* 165, 494 (1968).

9.  CHAPPERT, J., FRANKEL, R.B., MISETICH, A. and BLUM, N., *Phys. Rev.* 179, 578 (1969).

10. HAM, F.S., *Phys. Rev.* 160, 328 (1967).

11. RAHMAN, A., SINGWI, K.S. and SJOLANDER, A., *Phys. Rev.* 126, 986 (1962); ibid 126, 997 (1962).

12. MULLEN, J.G. and KNAUER, R.C., in *Mössbauer Effect Methodology*, Vol. 5, I. Gruverman, Ed. Plenum Press (1970), p. 197.

13. DEZSI, I., KESZTHELYI, L., MOLNAR, B. and POCS, L., in *Hyperfine Structure and Nuclear Radiations*, E. Matthias and D.A. Shirley, Eds. North-Holland (1968) p. 566.

14. NOZIK, A.J. and KAPLAN, M., *J. Chem. Phys.* 47, 2960 (1967)

15. PELAH, I. and RUBY, S., *J. Chem. Phys.* 51, 383 (1969).

16. SIMOPOULOS, A., WICKMAN, H.H., KOSTIKAS, A. and PETRIDES, D., These Proceedings,

17. ABRAGAM A. and PRYCE, M.H.L., *Proc. Roy. Soc.* A205, 135 (1951).

18. HIRST, L.L., *J. Phys. Chem. Solids* 31, 655 (1970).

19. HIRST, L.L., SEIDEL, E.R. and MÖSSBAUER, R.L., *Phys. Lett.* 29A, 673 (1969).

20. GABRIEL, H., *Phys. Stat. Sol.* 23, 195 (1967).

21. BLUME, M., *Phys. Rev.* 174, 351 (1968).

22. BOSSE, J., Diplomarbeit, Technischen Universität Carolo-Wilhelmina zu Braunschweig (1969).

23. BLUME, M. and TJON, J.A., *Phys. Rev.* 165, 446 (1968).

24. BRADFORD, E. and MARSHALL, W., *Proc. Phys. Soc.* 87, 731 (1966).

25. ZWANZIG, R., *J. Chem. Phys.* 33, 1338 (1960).

26. KUBO, R., in *Fluctuations, Relaxation, and Resonance in Magnetic Systems*, D. ter Haar, Ed. Oliver and Boyd (1962), p. 23.

27. WERTHEIM, G.K. and REMEIKA, J.P., *Phys. Lett.* 10, 14 (1964).

28. DOBLER, H., PETRICH, S., HUFNER, S., KIENLE, P., WEIDEMANN, W. and EICHER, H., *Phys. Lett.* 10, 319 (1964).

29. OFER, S., KHURGIN, B., RAKAVY, M. and NOWIK, I., *Phys. Lett.* 11, 205 (1964).

30. LANG, G., *Phys. Lett.* 26A, 223 (1968).

31. OOSTERHUIS, W.T., DEBENEDETTI, S. and LANG, G., *Phys. Lett.* 26A, 214 (1968).

32. VAN DER WOUDE, F. and DEKKER, A.J., *Phys. Stat. Sol. 13*, 181 (1966).

33. SUZDALEV, I.P., AFANAS'EV, A.M., PLACHINDA, A.S., GOLDANSKII, V.I. and MAKAROV, E.F., *Zh. Eksp. Teor. Fiz. 55*, 1752 (1968); *Soviet Phys. JETP 28*, 923 (1969).

34. HUFNER, S., WICKMAN, H.H. and WAGNER, C.F., *Phys. Rev. 164*, 247 (1968).

35. WICKMAN, H.H., *Phys. Lett. 31A*, 29 (1970).

36. WICKMAN, H.H. and WAGNER, C.F., *J. Chem. Phys. 51*, 435 (1969).

37. WICKMAN, H.H., KLEIN, M.P. and SHIRLEY, D.A., *Phys. Rev. 152*, 345 (1966).

38. VAN DER WOUDE, F. and DEKKER, A.J., *Solid State Comm. 3*, 319 (1965).

39. BOYLE, A.J.F. and GABRIEL, J.R., *Phys. Lett. 19*, 451 (1966).

40. GRODZINS, L., in *Hyperfine Structure and Nuclear Radiations*, E. Matthias and D.A. Shirley, Eds. North-Holland (1968), p. 607.

41. CHAMPENEY, D.C. and WOODHEMS, F.W.D., *J. Phys. B1*, 620 (1968), and references cited therein.

42. DEZSI, I., KESZTHELYI, L., MOLNAR, B. and POCS, L., *Phys. Lett. 18*, 28 (1965).

43. DILORENZO, J.V. and KAPLAN, M., *Chem. Phys. Lett. 3*, 216 (1969)

44. DAVIES, R.O. and JONES, B.O., *Adv. in Phys. 2*, 370 (1953).

45. SIMON, F.E. and LANGE, F., *Z. Phys. 38*, 227 (1926).

46. SUGISAKI, M., SUGA, H. and SEKI, S., *Bull. Chem. Sco. (Japan) 41*, 2586 (1968); ibid *45*, 2591 (1968).

47. PRYDE, J.A. and JONES, G.O., *Nature 170*, 685 (1952).

48. ROSS, R.T., *J. Chem. Phys. 42*, 3919 (1965).

DISCUSSION

H. DE WAARD: You have shown some cases where the agreement between theory and experiment on the relaxation behavior appears to be quite good. Of course there are also numerous cases where the agreement is considerably worse. I am thinking especially of the dilute ferric alum, which does not seem to have such a very complicated structure but where probably the $S_x$ and $S_y$ components (the transverse components of the spin) play a part and they cannot be neglected.

My question is, do you think the theory for this case, where you cannot assign an effective field, will be developed in the near future?

H.H. WICKMAN: I think that the situation is such that for any fluctuating classical field, electric or magnetic, the theories of Gabriel and Blume are adequate. Now when you couple the spin with the nucleus and ask for the energy spectrum and transition rates, the problem is much more difficult. However, I think, in principle the theory could be extended to that case. In the low field region, as in ferric alum, where you would need such a theory, you additionally have a complication of cross-relaxation. I think that this is exactly the region where you observe these effects. When you have cross-relaxation you have the possibility of extremely high order spin flips occurring. Now, even if you have the models, you are still faced with the calculation of these spin flips and this might be difficult still, but it would certainly take that as a minimum to interpret these ferric alum data.

H. DE WAARD: I would like to ask a second question,

about the drop in recoilless fraction, which you tentatively explain as a melting effect. There is an investigation by Dr. V.D. Gorovchenko (Kurchatov Institute, Moscow) of the recoilless fraction of $Pb(Fe_{0.5}Nb_{0.5})O_3$, with a Curie point at 95°C. There is a 12% dip in this fraction at the Curie point but it is very hard to believe that this is due to melting. Could the drop you find be due to an effect similar to that found by Gorovchenko?

H.H. WICKMAN: I did not have time to mention that there is also evidence from epr work that upon freezing of aqueous solutions, the metal ions segregate into glassy regions. For example, an epr study by Ross showed that frozen dilute $Mn^{2+}$ solutions gave an epr resonance similar to cases of strongly dipolar coupled metal ions in amorphous phases. The segregation of the metal ions is also consistent with Mössbauer results of Ruby and Pelah, who suggested a strong but long range coupling between metal ions in FAS. The long range coupling assumption is not necessary if the ions become associated during the freezing process. At the moment this event seems more likely than a strictly local process such as the one you mention.

J.I. VARGAS: Would you care to comment on the possible relaxation phenomena connected in angular correlation work with the feeding of the intermediate level by a gamma ray which is sufficiently energetic? I have in mind for instance, the case of $^{181}Ta$, where the recoil energy of the first gamma ray is about 0.03 eV. Now this is about 300°K, and presumably in the case of complexes you could excite optical modes which in the Mössbauer case you see as a change in the recoil free fraction but which, in the case of angular correlation, you could see as relaxation phenomena.

H.H. WICKMAN: What you describe may be possible in special cases. However, the lifetimes of the optical modes you describe, relative to the nuclear level lifetime, would be important. In many cases the lattice excitations could reach an equilibrium before γ-emission, in which cases the two techniques should show similar results.

J.I. VARGAS: As to relaxation during a phase transition, we have been doing some angular correlation work on ammonium heptafluohafniate crystals. If you look at the angular correlation above 0°C you have a purely dynamic coupling. If you go down through 0° your anisotropy drops drastically and then as you go further down to liquid nitrogen temperature it stays mixed static and dynamic. If you go up in temperature you find that you get a hysteresis - that is, at about 20°C you get the perturbation characteristic of the low temperature phase, the equivalent of the crystalline phase in your work.

E. BODENSTEDT: I should like to mention an extensive time differential perturbed angular correlation study which we have recently performed with $^{181}$Ta in frozen aqueous solutions. We used solutions of $HfF_4$ in 0.1N HF. We started at the temperature of liquid nitrogen where we saw a pure static interaction. We changed the temperature in steps of 10° and at the beginning we found no change of the time dependence and of the pure static interaction that was observed. When we reached a temperature of roughly -90°C an additional time dependent perturbation could be seen and when we went up 20° further (to approximately -70°C) the measurement revealed a pure time dependent perturbation without any static interaction. The attenuation was quite strong, which indicated a very short nuclear relaxation time. When we raised the temperature further in steps of 10°, we found a very strong temperature dependence of the nuclear relaxation time. Finally, at room temperature, we had practically no attenuation. Then we cooled the source again and as far as we could see, there was no hysteresis effect. We have as yet no good interpretation of these data.

H.H. WICKMAN: From what you have said, I do not believe there is any contradiction with the suggestion of an initially frozen glassy region. Upon heating the glass would melt and then recrystallize, with possibly quite different relaxation behavior in the two phases. After the irreversible glass to crystal transformation, no hysteresis effects would be expected.

J.C. WALKER:  Do I take it then, that you would even suggest that the changes in f that one occasionally sees in ferro-electric phase transitions might in some way be connected with some sort of melting and recrystallization, rather than a soft mode explanation?

H.H. WICKMAN:  No, I would not.  I think that whatever occurs in ferro-electric transitions  probably has its origins in non-amorphous effects.

# RELAXATION PHENOMENA DETECTED BY MÖSSBAUER EFFECT MEASUREMENTS IN FeNb$_2$S$_4$*

E. HERMON, H. RABBIE and S. SHTRIKMAN

*Department of Electronics, The Weizmann Institute of Science, Rehovot, Israel*

(Presented by E. Hermon)

Polycrystalline FeNb$_2$S$_4$ was synthesized by the ceramic method, and Mössbauer effect spectra taken at temperatures between 5 and 295°K using a $^{57}$Co in Pd source. A hyperfine splitting appeared at about 45°K, but no anomaly was observed at this temperature in either electrical resistivity or magnetic susceptibility measurements (Fig. 1). The Curie-Weiss law is obeyed above 70°K, and the Curie constant is 2.96 ± 0.01 emu-deg/mol.Oe, in good agreement with the theoretical value of 3.00 for divalent iron (S=2). Below 70°, the χ-T curve is somewhat flattened (Fig. 1), but no evidence of magnetic transition is observable. Neutron diffraction measurements also failed to show evidence of any magnetic transition down to 4.2°K [1]. The Mössbauer spectra (Fig. 2) exhibited temperature-dependent hyperfine structure. At 5°K a well resolved six line spectrum was observed. As the temperature increased, the lines broadened, the effective magnetic field remaining practically constant and a central paramagnetic peak appeared until at about 55°K, when an asymmetric two line

---

*This research has been sponsored in part by the Air Force Materials Laboratory Research and Technology Division AFSC, through the European Office of Aerospace Research, OAR, United States Air Force under contract F 61052 67 C 0040.

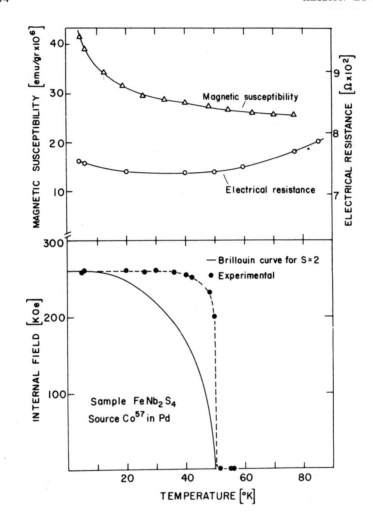

Fig. 1. Effective field, magnetic susceptibility and electric resistance as a function of temperature in $FeNb_2S_4$.

spectrum was visible. This asymmetry disappeared at about 70°K. The temperature dependence of the effective field determined from visual inspection of the spectra does not correspond at all closely to the Brillouin curve for S=2

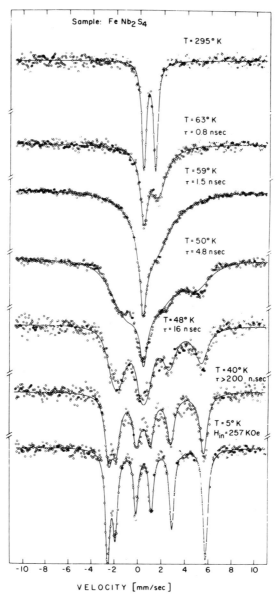

Fig. 2. Mössbauer spectra of FeNb$_2$S$_4$ at different temperatures. The solid lines are the computer fit for different relaxation times. The zero velocity is taken with respect to a $^{57}$Co in Pd source.

(Fig. 1). It was thus tempting to assume that a relaxation phenomenon was taking place for which this behaviour is characteristic [2,3]. Because of the striking similarity between the observed spectra and those calculated by Wickman et al. [2] on the basis of a simple model, a computer fit to their equation was attempted. The model adopted is that of an effective field which is constant in magnitude, but whose sign changes randomly with a relaxation time $\tau$. The results of the fit are shown in Fig. 2. Agreement is seen to be good in almost all cases. The small deviations for intermediate relaxation time values may be due to the simplicity of the model adopted. The relaxation time $\tau$ was found to vary exponentially with reciprocal temperature in the range 36-63° with an activation energy of about 350°K.

In $FeNb_2S_4$ the iron is divalent, as shown by the observed isomer shift of 0.9mm sec$^{-1}$ relative to stainless steel, the values of the Curie constant and of the effective field [4], and the stoichiometry of the compound, and so it would be expected that relaxation processes would be too rapid to be detectable by Mössbauer spectroscopy [2,5]. A single crystal grown by vapor transport technique (now in progress) will provide more data on the crystallography of the material, which at this stage are insufficient to permit detailed crystal field calculations. Also, it would be worthwhile to carry out specific heat studies to confirm that no magnetic transition occurs down to 4.2°K.

REFERENCES

1. SHAKED, H. Nuclear Center, Negev, Israel, private communication.

2. WICKMAN, H.H., KLEIN, M.P. and SHIRLEY, D.A. *Phys. Rev. 152*, 345 (1966).

3. BLUME, M. and TJON, J.A. *Phys. Rev. 165*, 446 (1968).

4. JOHNSON, C.E. *Symposia of the Faraday Society, 1*, London (1967).

5. CLARK, M.G. *J. Chem. Phys. 48*, 3268 (1968).

## DISCUSSION

M. KALVIUS: I am afraid that your last statement is not quite correct. There have been several observations of relaxation in divalent iron compounds. I remember that three or four years ago we measured pyroxenes, which are iron-magnesium silicates with divalent iron. They show relaxation spectra which are actually rather similar to yours. I think we had the same problem, that in the intermediate region we could not fit them because of electric field gradients. At that time the referee pointed out to us three or four earlier papers, which I cannot remember now. Of course it is quite well known relaxation phenomena in ferric compounds are more easily observed.

# HYPERFINE STRUCTURE IN f SHELL ELEMENTS*

I. NOWIK**

*Department of Physics, The Hebrew University, Jerusalem, Israel.*

ABSTRACT

In the following paper, some of the recent studies of hyperfine interactions in the rare-earth and actinide regions performed by the Jerusalem group, using the Mössbauer effect technique, are reviewed.

INTRODUCTION

The rare earth and actinide regions are very rich in isotopes having nuclear transitions suitable for Mössbauer effect experiments. This is a consequence of the prevalence of the low lying nuclear states near the nuclear ground state in the regions of strongly deformed nuclei. The chemical properties of the compounds of the 4f (rare earth) and 5f (actinide) transition elements

---

\* Invited paper.

\*\* On summer leave at the Technische Hochschule, Munich, Germany.

show some similarities. It is therefore interesting to compare the systematics of the hyperfine interactions in solid compounds of the rare earth and actinide elements. Though the rare earth region has been extensively studied by various groups since the early days of the Mössbauer technique, there are still many unexplored areas in nuclear and solid state physics open for further investigations using this technique. In this article some of the recent studies in the rare earth and actinide region performed by the Jerusalem group are presented.

The number of isotopes open for recoilless absorption studies in the 4f region has recently been increased to include several radioactive isotopes as absorbers, (e.g. the 2.6 year $^{147}$Pm, 87 year $^{151}$Sm and the 47 hr $^{153}$Sm) [1]. Some results on recoilless absorption measurements of the 91 keV gamma ray of $^{147}$Pm in various compounds will be presented here.

Until recently it was impossible to obtain well resolved Mössbauer spectra in any isotope of Gd so that no relevant information on the hyperfine interactions in Gd compounds from recoilless absorption measurements could be obtained; being an S-state ion, the hyperfine interactions in Gd compounds are of special interest. Recently quite well resolved Mössbauer spectra were obtained in $^{155}$Gd and the nuclear parameters of the 86.5 keV level of this isotope could be determined [2]. With the aid of these parameters, hyperfine interactions in some compounds could be analyzed - e.g. in GdIG. Relaxation phenomena in Gd compounds were observed. Mössbauer spectra of the 25.6 keV gamma ray of $^{161}$Dy in GdIG and DyIG indicate the existence of ionic level mixing in these compounds. The 25.6 keV gamma rays of $^{161}$Dy have been extensively used for Mössbauer studies since the earliest days of the Mössbauer effect - but the line widths obtained were always more than 10 times the natural line width ($2\Gamma$). Recently, much narrower lines were obtained using a $Dy^{4+}$ source. With these narrow lines, nuclear parameters and hyperfine fields can be determined with high accuracy.

Several recoilless absorption measurements in the

5f region (actinides) have been performed in recent years. As examples of measurements in this region — results obtained in some intermetallic Laves phase compounds of Np will be presented.

## 1. RECOILLESS ABSORPTION MEASUREMENTS OF THE 91 keV GAMMA-RAY OF $^{147}$Pm

Magnetically split recoilless absorption spectra in Pm are observed for the first time, enabling exact determination of the nuclear parameters of the 91-keV excited state of $^{147}$Pm. $^{147}$Pm is itself radioactive with a half life of 2.6 years. Once the nuclear parameters of this level are determined, the solid state properties of Pm in various compounds can be investigated.

Recoilless absorption spectra of the 91-keV gamma-ray transition from the 7/2+ excited state of $^{147}$Pm to the 5/2+ ground state were obtained using an absorber consisting of 100 mg/cm$^2$ $^{147}$Pm in the chemical form Pm$_2$O$_3$ with 50% impurities of Nd$_2$O$_3$ and Sm$_2$O$_3$. The beta activity due to the decay of $^{147}$Pm was $\sim$ 30 Ci. This was accompanied by considerable Bremsstrahlung. The recoilless absorption spectrum measured at 4.2°K using the above absorber and a $^{147}$Pm in NdAl$_2$ source is shown in Fig. 1. The absorber gives a single absorption line even at this low temperature, as was proved by carrying out a measurement with this absorber and a Nd$_2$O$_3$ source. The apparent line width was found to be 2.22 mm/sec, as compared with the natural line width ($2\Gamma_0$) of 1.16 mm/sec.

The spectrum was analyzed with a least-squares program. The best fit yielded the following parameters:

$g_0\mu_n H_{eff}/h$ = 2338 ± 22 MHz; $\frac{1}{4}eq_{eff}Q/h$ = 66 ± 30 MHz;

Q(91 keV)/Q(0) = 0.8 ± 0.4; g(91 keV)/g(0) = 1.925 ± 0.004.

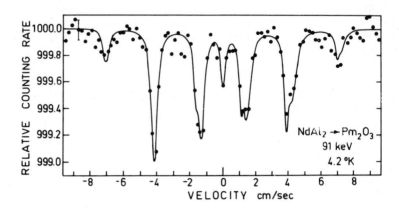

Fig. 1. Recoilless absorption spectrum of the 91 keV gamma rays emitted from a $^{147}$NdAl$_2$ source in a Pm$_2$O$_3$ absorber at 4.2°K. The solid line is the theoretical spectrum calculated using the parameters given in the text and assuming a small contribution due to Nd$_2$O$_3$ impurity in the source (the peak at zero velocity).

Assuming the value of 2.58 ± 0.07 n.m. for the magnetic moment of the ground state [3], the effective magnetic hyperfine field on the Pm nuclei in NdAl$_2$ is calculated to be 3.99 ± 0.16 MOe. This value is (95 ± 4)% of the saturated free ion value (4.2 MOe, calculated assuming $<1/r^3>$ equal to 6.18 $a_0^{-3}$ [4]).

The value of the magnetic moment of the 91-keV state, assuming the value of 2.58 ± 0.07 n.m. for the magnetic moment of the ground state [1], is 3.54 ± 0.10 n.m.

The 91 keV level in $^{147}$Pm has the same single-particle configuration as the ground state of $^{151}$Eu. The ground state of $^{147}$Pm and the first excited state of $^{151}$Eu are similarly related. The ratio of the g factor of the 5/2+ ground level in $^{151}$Eu to that of the 7/2+ first excited state was measured by Kienle [5] to be 1.88 as compared with our value of 1.92 in $^{147}$Pm.

Our value for the g factor ratio is not in agreement with the value 1.50 derived from the calculations of

# HYPERFINE STRUCTURE IN f SHELL ELEMENTS

Heyde and Brussard [6] using the unified model in intermediate coupling, nor is it in agreement with the value 2.22 derived from the calculations of Kisslinger and Sorensen [7] based on the theory of spherical nuclei with residual forces.

Similar measurements were carried out at 4.2 and 20°K using the same absorber and sources of $Nd_{0.2}Y_{2.8}Fe_5O_{12}$ and $Nd_{1.5}Y_{1.5}Fe_5O_{12}$. The results obtained with the $Nd_{0.2}Y_{2.8}Fe_5O_{12}$ source at 20°K is shown in Fig. 2. The spectra obtained with the first sample are consistent with the assumption that the magnetization is in the [111] direction. Least square fits yield the following values for $g_0\mu_n H_{eff}/h$ in the two $Pm^{3+}$ sites: 733 ± 22 MHz and 670 ± 17 MHz. The spectra obtained with the second sample are completely different from those obtained with the first sample and prove that the magnetization of $Nd_{1.5}Y_{1.5}Fe_5O_{12}$ is not in the [111] direction.

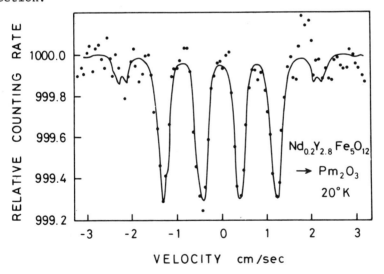

Fig. 2. Recoilless absorption spectra of the 91 keV gamma rays emitted from a $Nd_{0.2}Y_{2.8}Fe_5O_{12}$ source in a $Pm_2O_3$ absorber. The solid line is a theoretical spectrum with parameters obtained from a least square fit.

## 2. MÖSSBAUER AND NMR STUDIES OF $Gd_3Fe_5O_{12}$ (GdIG)

Mössbauer studies of the 86 keV transition of $^{155}Gd$ in GdIG were performed at 20°K using a $SmH_2$ source [2] (Fig. 3). The analysis of the spectrum yields the following parameters: $H_{eff}$ = 290 ± 20 KOe, $\frac{1}{4}eqQ_0/h$ = 20 ± 2 Mc/sec and $\eta$ = 0.7 ± 0.2. These results are in obvious contradiction with previously published NMR results [8]. 24 NMR lines are expected to be produced in the domains of GdIG. (The spin of the ground states of $^{155}Gd$ and $^{157}Gd$ is 3/2 and there are 2 magnetically inequivalent sites for $Gd^{3+}$ ions in GdIG.) Additional resonances are produced in the domain walls. We have calculated the expected NMR frequencies roughly and found that resonances should be observed at frequencies between 25 and 230 Mc/sec. NMR studies on GdIG were performed and in fact, in the whole region of 25 - 230 Mc/sec, resonances were observed.

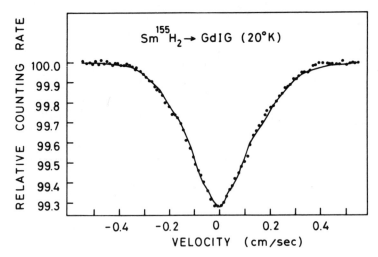

Fig. 3. Recoilless absorption spectrum of the 86.5 keV gamma-ray of $^{155}Gd$ in GdIG at 20°K (using a $SmH_2$ source). The solid curve is a theoretical fit to the spectrum.

## 3. RELAXATION EFFECTS IN $Gd_xY_{1-x}Cl_3 6H_2O$

Recoilless absorption measurements of the 86.5 keV gamma ray of $^{155}Gd$ have been used to study the hyperfine structure of $^{155}Gd$ in $Gd_xY_{1-x}Cl_3 6H_2O$ and in $SmCl_3 \cdot 6H_2O$ at 20°K. The results are shown in Fig. 4. The source used to study the $Gd_xY_{1-x}Cl_3 6H_2O$ absorbers was $SmH_2$ at 20°K [2] and the absorber used to study the $^{155}SmCl_3 \cdot 6H_2O$ was $GdRh_2$ at 80°K, which gives a narrow unsplit absorption line [2]. The spectra show clearly concentration dependent relaxation effects, indicating that the spin-spin relaxation rates are comparable to the nuclear Larmor frequency. Assuming that for the $Gd^{3+}$ ions in these samples the magnetic effective field is $\sim$ 300 KOe and that it fluctuates along the axis of the electric field gradient, one can analyze the spectra following the theory given by Wegener [9]. The theoretical Mössbauer spectra are quite similar to the experimental spectra. The analysis yields the following relaxation times: $4.10^{-10}$ sec for $x = 1.0$, $4.10^{-9}$ sec for $x = 0.1$ and $4.10^{-8}$ for $x = 0.01$. The theory is not applicable for the the spectrum obtained with the $^{155}SmCl_3 \cdot 6H_2O$ source.

## 4. HYPERFINE STRUCTURE OF $^{161}Dy$ IN GdIG, DyIG AND SmIG

The recoilless absorption spectrum in $DyF_3$ at room temperature of the 25.6 keV gamma ray of $^{161}Dy$ emitted from a $^{161}Tb$ in GdIG source at 4.2°K is shown in Fig. 5(a). The absorption spectrum in DyIG at 4.2°K of the 25.6 keV gamma rays emitted from a $^{161}Tb$ in $GdF_3$ source at room temperature is shown in Fig. 5(b). In GdIG and DyIG the magnetization is in the [111] direction of the unit cell. It is therefore quite surprising that the two spectra are so different. In addition the two spectra are inconsistent with the effective magnetic field approximation, assuming two sites for the $Dy^{3+}$ ions. A possible explanation for the spectra is that the first excited ionic level lies close to the ionic ground level. Crude calculations show that the spectrum in GdIG may be explained assuming a separation of 1.4°K between the two lowest levels, whereas, in order to explain the DyIG spectrum, a separation of about 0.3°K has to be assumed.

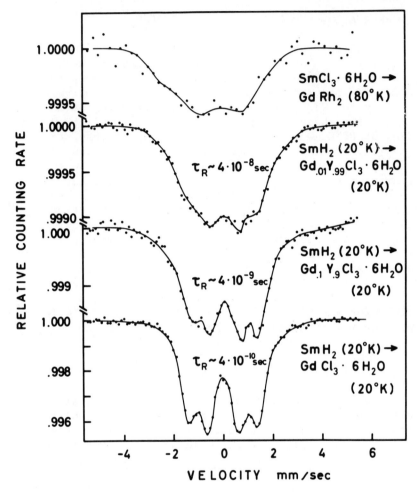

Fig. 4. Recoilless absorption spectra of the 86.5 keV gamma ray of $^{155}$Gd.

In Fig. 5(c) the spectrum of $^{161}$Dy in SmIG at 4.2°K is shown. The spectrum obtained is different from the spectra in GdIG and DyIG. This is not surprising in light of the fact that in SmIG the magnetization is in the [110] direction and the Dy-Fe exchange interaction is anisotropic.

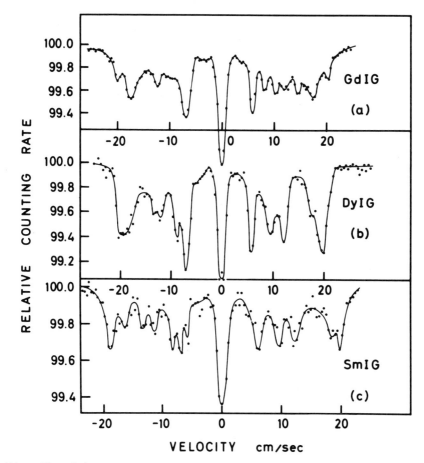

Fig. 5. (a) Recoilless absorption spectrum in $DyF_3$ at 300°K of the 25.6 keV gamma ray of $^{161}Dy$ emitted from a $^{161}Tb$ source in GdIG at 4.2°K.
(b) Recoilless absorption spectrum in DyIG at 4.2°K of the 25.6 keV gamma ray emitted from a $^{161}Tb$ source in $GdF_3$ at 300°K.
(c) Recoilless absorption spectrum in $DyF_3$ at 300°K of the 25.6 keV gamma ray emitted from a $^{161}Tb$ source in SmIG at 4.2°K.

## 5. RECOILLESS ABSORPTION MEASUREMENTS OF THE 25.6 keV GAMMA RAY OF $^{161}$Dy EMITTED FROM $^{161*}$Dy NUCLEI IN Dy$^{4+}$ IONS

We have carried out recoilless absorption measurements of the 25.6 keV gamma rays of $^{161}$Dy using a $^{161}$Tb($^{161}$Dy) in $CeO_2$ source. The source was prepared by neutron irradiation of a few mgs of $Gd_2O_3$ enriched in $^{160}$Gd. The $^{161}$Tb activity was separated from the gadolinium by ion exchange and precipitated in 2 gms of cerium oxalate. The precipitate was fired at 850°C for 24 hours. The central part of the absorption spectrum in a $^{161}$Dy metal absorber of the 25.6 keV gamma rays emitted from the $^{161}$Tb in $CeO_2$ source at room temperature is shown in Fig. 6. The spectrum consists of 4

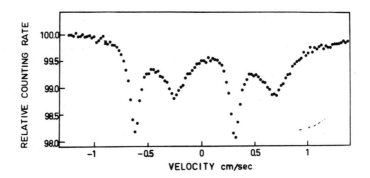

Fig. 6. Central part of the recoilless absorption spectrum in a metallic $^{161}$Dy absorber at 4.2°K of the 25.6 keV gamma rays emitted from a $^{161}$Tb in $CeO_2$ source at room temperature.

lines: Two narrow lines with a width at half height of 1.2 mm/sec are identified as absorption lines of the gamma rays emitted from $^{161*}$Dy in a Dy$^{4+}$ ion, whereas, the two broad lines (∿ 3.7 mm/sec) are identified as absorption lines of gamma rays emitted from $^{161}$Dy* in a Dy$^{3+}$ ion. The isomer shift between the 25.6 keV gamma rays emitted from Dy$^{4+}$ ions to those emitted from Dy$^{3+}$ ions was determined to be (3.71 ± 0.04) mm/sec. Henning et al [10] have previously reported an isomer shift of about 6.5 mm/sec for the 25.6 keV gamma-rays of $^{161}$Dy between $Dy^{3+}$ and $Dy^{2+}$ ions. The 25.6 keV transition of $^{161}$Dy is the only transition in the rare-earth region for which absorption measurements were carried out for 3 different valence states. As $CeO_2$ is not a perfect ionic compound, it is not clear yet whether the isomer shift measured in the present case represents the difference between $Dy^{3+}$ and $Dy^{4+}$ in pure ionic compounds.

The absorption spectrum in a $^{161}$Dy metallic absorber at 4.2°K of the 25.6 keV gamma rays emitted from the $^{161}$Tb in $CeO_2$ source at 4.2°K is shown in Fig. 7. Only

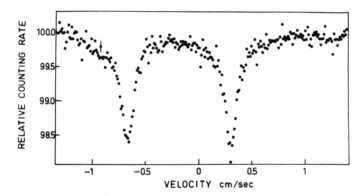

Fig. 7. Central part of the recoilless absorption spectrum in a metallic $^{161}$Dy absorber at 4.2°K of the 25.6 keV gamma rays emitted from a $^{161}$Tb in $CeO_2$ source at 4.2°K.

two absorption lines corresponding to the four-valent emission line are seen. The "trivalent" emission line is split at 4.2°K and its absorption lines cannot be detected in Fig. 7. The separation between the two lines in Fig. 7 was accurately measured and, relying on the value of (831.1 ± 1)Mc/sec determined in NMR measurements [11] for $g_0\mu_n H_{eff}$ of the ground level of $^{161}$Dy in Dy metal at 4.2°K, the ratio of the g-factors of 25.6 keV level and the ground level of $^{161}$Dy was found to be 1.241 ± 0.004.

## 6. MÖSSBAUER EFFECT STUDIES OF CUBIC LAVES-PHASE COMPOUNDS OF Np*

Recoilless absorption measurements of the 59.6 keV gamma rays of $^{237}$Np in $NpFe_2$, $NpCo_2$, $NpNi_2$ and $NpIr_2$ were performed at various temperatrues. The source used was 5% $^{241}$Am embedded in thorium. The absorption spectrum in $NpFe_2$ at 4.2°K is shown in Fig. 8. The results of the analysis of the spectra are summarized in Table I. The table also includes previously published results on $NpAl_2$ [12, 13]. The systematic behavior of $H_{eff}$ in these compounds is different from the corresponding behavior of the rare-earth Laves-compounds. In the case of the rare earths, the hyperfine magnetic fields are usually close to the free ion values at magnetic saturation due to the fact that most of the hyperfine field is produced by the well screened 4f electrons. In the actinide compounds the hyperfine fields are more effected by crystalline fields and conduction electron contributions.

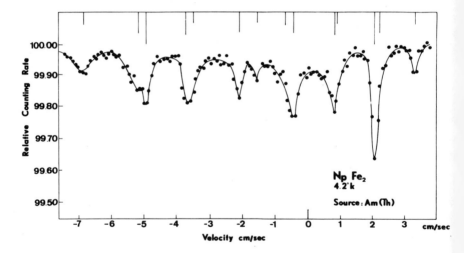

Fig. 8. Recoilless absorption spectrum of the 59.6 keV gamma ray of $^{237}$Np, obtained with a 5% $^{241}$Am in Th source and a $NpFe_2$ absorber at 4.2°K.

---

* This work has been performed at the Nuclear Research Center, Negev, Israel.

TABLE 1 - Hyperfine Structure Constants, Isomer Shift and Curie Temperatures of some Np Laves Phase Compounds. The Isomer Shifts are given with respect to $NpO_2$.

| | T°K | $g_0\mu_n H_{eff}$ cm/sec | $H_{eff}/H_{eff}(NpAl_2)$ | 1/4 eqQ cm/sec | Isomer Shifts cm/sec | $T_c$ °K |
|---|---|---|---|---|---|---|
| $NpNi_2$ | 4.2 | 4.5 ± 0.2 | 0.84 ± 0.10 | 0.03 ± 0.10 | -1.10 ± 0.10 | 19 ± 2 |
| $NpFe_2$ | 4.2 | 3.00 ± 0.02 | 0.562 ± 0.010 | +0.27 ± 0.03 | -1.80 ± 0.04 | >300 |
| $NpCo_2$ | 1.7 | 1.65 ± 0.16 | 0.31 ± 0.03 | — | -1.70 ± 0.05 | 8.0 ± 0.5 |
| $NpIr_2$ | 1.7 | 2.07 ± 0.09 | 0.388 ± 0.017 | — | -0.49 ± 0.02 | 5.5 ± 0.5 |
| $NpAl_2$ | 4.2 | 5.34 ± 0.05a) | 1.0 | -0.015 ± 0.005b) | 0.57 ± 0.05a) | 55.8 ± 0.1a) |

a) From Ref. 11.

b) From Ref. 12.

The larger magnetic field and smaller quadrupole interaction in $NpAl_2$ as compared with the smaller magnetic field and larger quadrupole interaction in $NpFe_2$, tend to show that the field at the Np nucleus caused by the polarization of the 5f electrons is larger in $NpFe_2$ than $NpAl_2$ and that the smaller field found in $NpFe_2$ results from a contribution of opposite sign to the magnetic field caused by the polarized conduction electrons. This contribution is much larger in $NpFe_2$, where the conduction electrons are polarized by the iron ions, than in $NpAl_2$.

The isomer shifts obtained in all Np intermetallic Laves phase compounds investigated, are between the isomer shifts obtained for other non-metallic compounds containing Np ions in a +4 and +5 valence state [14]. This tends to show that the valence of the Np is 4 in these compounds, and the additional shift results from the contribution of the conduction electron density at the nucleus. This contribution is the smallest in $NpAl_2$ and the largest in $NpFe_2$.

ACKNOWLEDGMENT

I am very grateful to S. Ofer and R. Bauminger for allowing me to present some of their experimental results prior to publication, and for their help in preparing the manuscript.

REFERENCES

1. BOYLE, A.J.F. and PERLOW, G.J., in *Mössbauer Effect Methodology* (I.J. Gruverman, ed.), Vol. 5, 183 (1970).

2. BAUMINGER, E.R., FROINDLICH, D., MUSTACHI, A., NOWIK, I., OFER, S., SAMUELOV, S., *Phys. Lett. 30B*, 531 (1969).

3. READER, J., *Phys. Rev. 141*, 1123 (1966).

4. BLEANEY, J., *Proc. 3rd Intern. Symp. Quantum Electron.*, Paris 1963, p. 595.

5. KIENLE, P., *Rev. Mod. Phys. 36*, 372 (1964).

6. HEYDE, K., BRUSSAARD, P.J., *Nucl. Phys. A104*, 81 (1967).

7. KISSLINGER, L.S. and SORENSEN, R.A., *Rev. Mod. Phys. 35*, 853 (1963).

8. LE DANG KHOI, *Phys. Lett. 28A*, 671 (1969).

9. WEGENER, H., *Z. Physik 186*, 498 (1965).

10. HENNING, W., KAINDL, G., KIENLE, P., KORNER, M.J., KULZER, H., BEHM, K.E., *Phys. Lett. 28A*, 209 (1968).

11. KOBAYASHI, S., SANO, N. and ITOH, J., *J. Phys. Soc. Japan 2*, 1456 (1966).

12. STONE, J.A. and PILLINGER, W.L., *Phys. Rev. 165*, 1319 (1968).

13. DUNLAP, B.D., BRODSKY, A.B., KALVIUS, G.M., SHENOY, G.K. and LAM, D.J., *J. Appl. Phys. 40*, 1495 (1969).

14. DUNLAP, B.D. KALVIUS, G.M., RUBY, S.L., BRODSKY, M.B. and COHEN, D., *Phys. Rev. 171*, 316 (1968).

DISCUSSION

J.C. WALKER: I would like to make a comment and ask a question. Amusingly enough, we also did the $^{147}$Pm in $NdAl_2$, and the agreement in g values is very good - we got 1.917 and you got 1.924, so it shows that things are correct.

You were saying that in most of the rare earths you always found the free ion hyperfine field. This is, conspicuously, not the case in neodymium compounds the results of which are reported in another contribution to this conference.

I. NOWIK: I was very careful to say 'almost'.

J.C. WALKER: How many parameters did you use in analyzing your gadolinium iron garnet data?

I. NOWIK: We used three parameters, but the data is not sensitive to the third parameter, η, so actually there are only two left.

J.C. WALKER: I assume that you put in the angle between the electric field gradient and the magnetic field from other considerations?

I. NOWIK: Yes.

J.C. WALKER: Because with such unresolved spectra - at least, this has been our experience - it is rather difficult, no matter how few parameters, to determine them very well.

I. NOWIK: The angle is well known from geometrical considerations. We know that the hyperfine field is along the magnetic moment and the z axis of the local orthorhombic coordinate system of the rare-earth ion.

W. HENNING: Regarding the line narrowing of the 105 keV state in $^{155}$Gd: As far as I remember, Blumberg and Persson have also observed this and they attribute it somehow to the different feeding in the nuclear decay. What is your finding about this?

I. NOWIK: The linewidth at half height is 60% of the natural linewidth. It means that it is narrower than the natural linewidth obtained from lifetime measurements. This was explained in the paper by Blumberg et al. as an after-decay local mode relaxation effect - and the meaning of this actually is that the line is not a Lorenzian line but has a shape where at half height it is narrower but has a long tail. So it is only a difference in shape which, if the only quantitative measurement is the linewidth at half height, reveals itself as a narrower line.

M. KALVIUS: I am very happy that after many years of competing in the rare earths, we can compete now in the actinides as well. I would like to make a few remarks concerning the actinides. There are two more isotopes which are quite promising - one is $^{243}$A and the other is $^{231}$Pa. Further, in the actinides the core polarization fields are considerably larger than in the rare earths; an estimate of the core field for Am gives about 500 kOe, instead of only a few kOe in the analogous rare earth. So this contribution cannot be neglected here.

Next one has to be rather careful with the $<r^{-3}>$ because relativistic effects play an important role and I would like to point out that recently the Los Alamos group has published an extensive calculation on $<r^{-3}>$ from relativistic Dirac-Fock calculations.

The last comment is that the Sternheimer antishielding factor is also considerably larger in the actinides and we have made an estimate on it which gives about 0.35. This is also considerably larger than in the rare earths. Finally, when I collected data on the isomer shifts in dysprosium, Dr. R.L. Cohen from Bell Labs. gave me his numbers for shifts obtained with $3^+$, $4^+$ and even $5^+$ Dy ions and I think this fitted rather well. I plotted it and it came out that it was not so different from the neptunium case. Would Dr. Cohen like to comment on it?

R.L. COHEN: I wanted to return to this question of the dysprosium isomer shifts. We have observed, in the same way as you have, from a radioactive source, metastable states in which we see not only $Dy^{4+}$ but a line that is apparently identifiable as $Dy^{5+}$. So we have not only the $4^+$, $3^+$ shift but the $5^+$, $4^+$ shift as well. We have also tried, and partly duplicated the results of the Munich group on the $2^+$, $3^+$ shift; this is still at an early stage, but it looks like the isomer shifts are pretty much the same between all of these different valence states. Maybe you, Dr. Henning, would like to comment - I don't know if the isomer shift between the divalent and trivalent dysposium was 6.5 mm/sec. I thought it was somewhat smaller than that.

W. HENNING: It was about that.

F.E. WAGNER: I would just like to make a comment on these isomer shifts. We have found about the same thing in praseodymium as you did in dysprosium - trivalent and tetravalent Pr form stable compounds and one can look at

tetravalent Pr in fluorides and in oxides. The shifts
between trivalent and tetravalent fluorides are about
twice as large as between the tetravalent oxide and the
trivalent fluoride. So that is about the same as you observe
in dysprosium.

J.C. TRAVIS: Just continuing Dr. Walker's point on the
technology of curve fitting: Our experience, at the
National Bureau of Standards, of how to constrain curve
fitting of unresolved spectra has been that line intensity
problems - such as asymmetries and thickness broadening -
make the hyperfine parameters just unmeasureable in cases
like this.

I. NOWIK: Well, I said we used three parameters for the
fitting and I said that the spectrum is not so sensitive
to the third parameter. But not only is it not sensitive
to the third parameter, but also the quadrupole interaction
has been measured independently, in a gadolinium
gallium garnet which is paramagnetic and not ferromagnetic,
and you get the same value for the electric field gradient.
So, it reduces another parameter.

P. KIENLE: Did you also measure isomer shifts in the case
of promethium? That should be quite interesting for the
systematics of the nuclei in this region.

I. NOWIK: The isomeric shift between Pm in $NdAl_2$ and the
oxide is very small and hard to measure; I cannot give
you the exact figure.

# MAGNETIC HYPERFINE FIELDS ON HOLMIUM AND LUTETIUM IN Ho-Lu ALLOYS

W. SCHOTT

*Institut für Angewandte Kernphysik, Kernforschungszentrum Karlsruhe, Germany, and Brookhaven National Laboratory, Upton, New York*

V.L. SAILOR

*Brookhaven National Laboratory, Upton, New York*

(Presented by W. Schott)

ABSTRACT

From transmission measurements of polarized neutrons with polarized targets, the hyperfine fields in 3 Ho-Lu alloys were determined. The alloys contained 19, 40 and 81 atomic percent holmium. On holmium the measured fields were +5041 ± 187, +6128 ± 227, and +6824 ± 183 kOe, on lutetium +22 ± 4, +128 ± 25, and +787 ± 129 kOe, respectively.

By measuring the transmission effect $\varepsilon = (T_P - T_A)/(T_P + T_A)$ at a neutron resonance energy of a special nucleus of the sample, it is possible to determine the polarization of that nucleus if the spin of the resonance is known. ($T_{P,A}$ is the sample transmission of neutrons polarized paralled and antiparallel to an externally applied magnetic field.) From the nuclear polarization and the measured sample temperature the hyperfine field at the nucleus can be determined. We

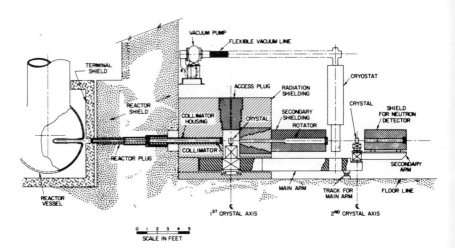

Fig. 1. Neutron spectrometer for transmission measurements of polarized neutrons with polarized targets.

used the resonances at 3.92 eV of $^{165}$Ho and at 0.142 eV of $^{176}$Lu to measure the fields on holmium and lutetium. Fig. 1 shows the transmission spectrometer for polarized neutrons and polarized targets at the high flux beam reactor in Brookhaven National Laboratory. Low temperatures down to 30 mK were obtained by adiabatic demagnetization of a paramagnetic Cr-alum salt or by dilution refrigation. Using a superconducting magnet a constant field of 37.2 kOe was applied to the sample. If several nuclei $\alpha$ with abundances $F_\alpha$ and polarizations $f_{N_\alpha}$ contribute to the transmission effect, we get [2]

$$\varepsilon = -\tfrac{1}{2}(1 + \phi) f_n^0 \times \frac{\int_0^\infty R(E-E')\tau \sinh(\kappa t) \, e^{-\delta t} \, dE'}{\int_0^\infty R(E-E') \, e^{-\delta t} \{\cosh(\kappa t) - [\tfrac{1}{2}(1-\phi) \, f_n^0 - \nu]\sinh(\kappa t)\} \, dE'}$$

$$\kappa = [(N \sum_{i,\alpha} F_\alpha f_{N_\alpha} \rho_{i,\alpha} \sigma_{i,\alpha})^2 + D^2]^{\tfrac{1}{2}}$$

$$\tau = \frac{1}{\kappa} (N \sum_{i,\alpha} F_\alpha f_{N_\alpha} \rho_{i,\alpha} \sigma_{i,\alpha})$$

$$\nu = \frac{D}{\kappa}, \qquad \delta = \sum_{i,\alpha} F_\alpha \sigma_{i,\alpha}$$

$f_n^0$ is the neutron polarization, N·t is the number of nuclei per sample area, $\sigma_{i,\alpha}$ is the total cross section of the resonance i of the nucleus $\alpha$, $\delta_{i,\alpha}$ is a weighting factor which is $I_\alpha/(I_\alpha + 1)$ or $-1$ depending on whether the spin of the resonance i is $I_\alpha + \frac{1}{2}$ or $I_\alpha - \frac{1}{2}$, respectively, $\phi$ is the beam flipping efficiency and D is the depolarization factor, R(E) is the resolution function of the spectrometer.

Using these formulae the nuclear polarization of holmium and lutetium was obtained from the $\varepsilon$-values which were measured at the resonance energies 3.92 and 0.142 eV taking into account small contributions of other resonances to the transmission effect. In the case of holmium the considered resonances and their spins [3] were: $^{175}$Lu (2.57 eV, 7/2 + 1/2), $^{165}$Ho (3.92 eV, 7/2 + 1/2), $^{176}$Lu (4.37 eV, 7 + 1/2), $^{175}$Lu (4.80 eV, 7/2 + 1/2), $^{175}$Lu (5.22 eV, 7/2 − 1/2); in the case of lutetium: $^{176}$Lu (0.142 eV, 7 − 1/2), $^{176}$Lu (1.56 eV, 7 + 1/2), $^{175}$Lu (2.576 eV, 7/2 + 1/2), $^{165}$Ho (bound resonance [4]).

Since the sign of $\varepsilon$ is determined by the sign of $\rho$, i.e. by the spin of the resonance and the sign of $f_N$, the sign of the hyperfine field is obtained in this experiment too.

We neglect the electric hyperfine interaction in both cases. The Ho-nuclei are entirely polarized at the lowest temperatures along the directions of the hyperfine fields. Having obtained $f_{N_{Ho}}$ at different sample temperatures T, we get a magnetization constant $k_m$ and the hyperfine coupling constant A from a fit of the function $f_{N_{Ho}} = k_m B(A/T)$ to the measured $f_{N_{Ho}}$ (1/T) values (B is the Brillouin function). $k_m$ is smaller than one because the sample is a polycrystal and is not magnetized to saturation. Therefore, only a fraction $k_m$ of the holmium magnetic moments and nuclear moments can be polarized along the external magnetic field. The hyperfine fields

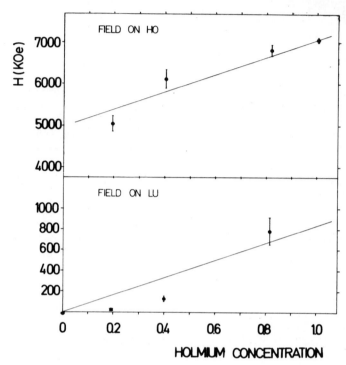

Fig. 2. Hyperfine fields on Holmium and Lutetium in Ho-Lu alloys.

on lutetium are much smaller, therefore only about 10 percent of the Lu-nuclei are polarized at the lowest temperatures. In this region the Brillouin function is linear in 1/T. Therefore a hyperfine field on lutetium is obtained from the slope of the $f_{N_{Lu}}$ (1/T)-curve and from $k_m$ which is known from the holmium measurement. The hyperfine fields were calculated using $\mu_{165_{Ho}}$ = +4.12 nm and $\mu_{176_{Lu}}$ = +3.18 nm [5].

Fig. 2 shows the hyperfine fields (corrected for local fields) versus Ho-concentration. Included are H-values for Ho-metal [6] and Lu-metal [3]. In both cases there is no linear dependence on the Ho-concentration. If we try to fit a straight line to the points the slopes are different by at least a factor of 2.5.

Only the contribution to the hyperfine fields which are caused by conduction electron polarization should depend on the Ho-concentration. The fields on lutetium are caused purely by conduction electron polarization of the surrounding Ho-ions. These fields are positive, which is in agreement with a rough spin density calculation we made, using the form factor of $Ho^{3+}$ -ions as an exchange integral $J(Q)$ between the spins of the Ho-ions and the conduction electron spins. We used for the spin density [7]

$$\rho(r) \sim \frac{1}{r} \int_0^\infty \sin(Qr) \; J(Q) \; \chi(Q) \; QdQ$$

A positive spin density results at the nearest neighbor site causing a positive field at the Lu-nucleus.

Zmora et al. [8] obtained a linear behavior of the fields on Gd and Lu in Gd-Lu alloys over the whole range of Gd-concentration. A nonlinear behavior of the fields can be caused by a change of the structure of the conduction electrons depending on the Ho-concentration. Such a change could influence the magnetic structure too. There is an indication that the magnetic structure of the Ho-Lu alloys depends on the Ho-concentration. For 80% Ho-concentration the sample is ferromagnetic like the pure metal; $k_m = 0.67$ is compatible with the polycrystal nature of the sample and with the fact that the sample is not magnetized to saturation. The depolarization D = 2.43 $cm^{-1}$ is reasonable for ferromagnetic holmium [9]. At about 60% Ho-concentration the magnetic structure seems to change, because $k_m$ drops to 0.15 and D to about 0.2 $cm^{-1}$. It is possible that the sample becomes antiferromagnetic at about 60% Ho-concentration. For smaller Ho-concentrations $k_m$ rises again, because the antiferromagnetic structure is more and more disturbed. For a Ho-concentration of 20 percent, $k_m$ is 0.39. The depolarization remains very low because of the small amount of magnetic ions.

The authors are grateful to Dr. G. Brunhart and Dr. R.I. Schermer for their help concerning the experiment, further to Mr. W. Drexel for writing the computer program for the spin density calculations. One of us (W.S.) would like to express his appreciation for the kind hospitality of the Nuclear Cryogenic Group at BNL during his stay as a visiting physicist.

REFERENCES

1. BRUNHART, G., POSTMA, H. and SAILOR, V.L., *Phys. Rev. 137B*, 1484 (1965).

2. SHORE, F.J., REYNOLDS, C.A., SAILOR, V.L. and BRUNHART, G., *Phys. Rev. 138B*, 1361 (1965).

3. The spins of the lutetium resonances were taken from H. Postma et al., to be published.

4. SCHERMER, R.I., *Phys. Rev. 136B*, 1285 (1964).

5. FULLER, G.H. and COHEN, V.W., *Nuclear Data Tables A5*, 433 (1969).

6. GORDON, J.E., DEMPSEY, C.W. and SOLLER, T., *Phys. Rev. 124*, 724 (1951) and [1].

7. WATSON, R.E. and FREEMAN, A.J., *Phys. Rev. 152*, 566 (1966).

8. ZMORA, H., BLAU, M. and OFER, S., *Phys. Lett. 28A*, 668 (1969).

9. POSTMA, H., MARSHAK, H., SAILOR, V.L., SHORE, F.J. and REYNOLDS, C.A., *Phys. Rev. 126*, 979 (1962).

DISCUSSION

R. L. COHEN: Regarding the analysis of this data, it seems to me that you are taking a disordered alloy (it is an ordered crystal structure, which is disordered as far as having holmium or lutetium in a particular lattice site), and you implicitly assume in the analysis that there is a unique hyperfine field, the same for all the holmiums or the same for all the lutetiums. That is in principle really not a realistic assumption when you have a disordered material.

W.K. SCHOTT: I assume that the hyperfine field on the holmium is directed along the magnetic moment of the holmium and only a special component of the magnetic moment and the hyperfine field is directed along the external field. I observe an average hyperfine field, of course.

R.L. COHEN: What I am concerned about is that you assume that you have the same hyperfine field for each holmium atom - holmium atoms are in many different surroundings, depending on the number of holmium nearest neighbors.

W.K. SCHOTT: Yes, that is correct.

# ON THE MAGNETIC FIELD AND CHARGE DENSITY OF CONDUCTION ELECTRONS AT NUCLEI IN RARE EARTH METALS

W. HENNING, G. BÄHRE and P. KIENLE

*Physik-Department der Technischen Universität München, Munich, Germany*

(Presented by W. Henning)

For further investigation of the charge density and magnetic field of conduction electrons (C.E.) at nuclei of rare earth (R.E.) metals, we have determined isomer shifts and magnetic splitting of R.E. nuclei embedded in various 4f-metals, using the Mössbauer effect. The method of measuring hf parameters in neighboring metals seems favorable for the following reasons. a) Highly dilute solutions are obtained by exposing the metal to some nuclear reaction and using the sample as an MB source. b) A great number of MB transitions exist, owing to the high density of low lying Nilsson states and rotational states in this mass region. Thus nearly all rare earth metals may be studied (Fig. 1). c) The very close chemical and crystalline properties introduce only small changes in physical properties such as Wigner-Seitz radius and band structure, allowing a more controlled study on their influence.

Table I gives results on isomer shift measurements. Column 5 gives the ratio of the deduced C.E. density D(CE) at the nucleus relative to the electron density difference between trivalent and divalent rare earth ions. In column 6 the relative change in C.E. density at the metallic ion in neighboring metals compared to its own metal is listed. The following conclusions are drawn:

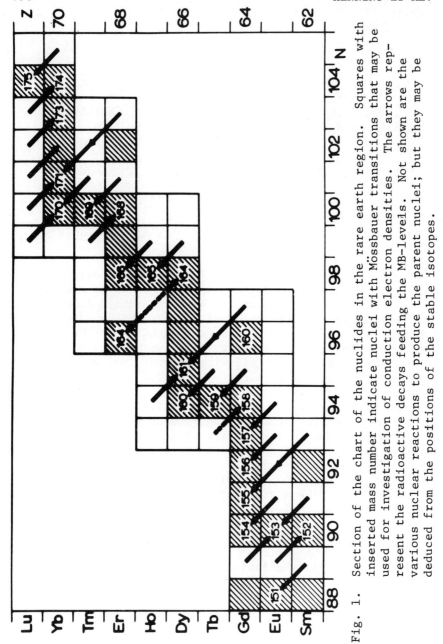

Fig. 1. Section of the chart of the nuclides in the rare earth region. Squares with inserted mass number indicate nuclei with Mössbauer transitions that may be used for investigation of conduction electron densities. The arrows represent the radioactive decays feeding the MB-levels. Not shown are the various nuclear reactions to produce the parent nuclei; but they may be deduced from the positions of the stable isotopes.

TABLE I — Isomer shifts and deduced conduction-electron densities for several rare earths metals.

| Transition | Source | Absorber | Isomer shifts (mm. sec$^{-1}$) | $\dfrac{D(CE)}{D(RE^{3+} - RE^{2+})}$ | $\dfrac{D(CE)_{ngb} - D(CE)_{met}}{D(CE)_{met}}$ | $\left(\dfrac{\Delta V}{V}\right)$ met. rad. |
|---|---|---|---|---|---|---|
| $^{153}$Eu 103 keV | Sm$_2$O$_3$ | Eu-metal | +10.3 ± 0.3 | +0.41 ± 0.03 | -- | -- |
|  | Eu in Gd-metal | Eu$_2$O$_3$ | -9.2 ± 0.2 | +0.46 ± 0.03 | +0.16 ± 0.06 | -0.32 |
|  | Eu in Sm-metal | Eu$_2$O$_3$ | -7.8 ± 0.1 | +0.55 ± 0.04 | +0.36 ± 0.06 | -0.32 |
| $^{155}$Gd 87 keV | SmF$_3$ | Gd-metal | -0.66 ± 0.04 | +0.40 | -- | -- |
|  | Gd in Sm-metal | GdF$_3$ | +0.51 ± 0.06 | +0.32 | -- | +0.001 |
|  | Gd in Sm-metal | Gd-metal | -0.12 ± 0.06 | -- | -0.20 ± 0.08 | -- |
| $^{161}$Dy 26 keV | GdF$_3$ | Dy-metal | +2.7 ± 0.1 | +0.43 ± 0.03 | -- | -- |
|  | Dy in Gd-metal | DyF$_3$ | -2.25 ± 0.05 | 0.36 ± 0.02 | -- | +0.03 |
|  | Dy in Gd-metal | Dy-metal | +0.5 ± 0.2 | -- | -0.17 ± 0.04 | -- |
| $^{170}$Yb 84 keV | Yb in Tm-metal | Yb$_2$S$_3$ | +0.06 ± 0.04 | +0.50 ± 0.12 | -- | -- |
|  | Yb in Tm-metal | Yb-metal | -0.12 ± 0.02 | +0.75 ± 0.12 | +0.5 ± 0.3 | -0.30 |
| $^{174}$Yb 76 keV | Yb in Lu-metal | Yb-metal | -0.26 ± 0.07 | 1.3 ± 0.4 | +1.6 ± 0.8 | -0.37 |

TABLE II - Results of Mössbauer measurements for the magnetic field at Eu-nuclei in Gd-, Sm-, and Eu-metal.

| γ-Transition and γ-source | Absorber | max. effect | magn. field (kGs) |
|---|---|---|---|
| $^{153}$Eu 103 keV Gd-metal | $Eu_2O_3$ 200 mg/cm$^2$ | $7 \times 10^4$ | $-290 \pm 35$ |
| $^{153}$Eu 103 keV Sm-metal | $Eu_2O_3$ 200 mg/cm$^2$ | $6 \times 10^4$ | $-310 \pm 25$ |
| $^{151}$Eu 21.7 keV $Sm_2O_3$ [1] | Eu-metal | -- | $-265 \pm 5$ |

a) In the pure metal D(CE) stays about constant from Eu to Yb. Thus, crudely, the number of 6s C.E. is the same for divalent and trivalent metals.

b) The change of D(CE) in neighboring metals is positive and largest for the divalent ions diluted in trivalent metals. This most probably results from the compression of the large divalent ions, i.e. a reduction of the Wigner-Seitz radius. For comparison, in column 7 the relative volume changes ΔV/V are calculated from the metallic radii.

The decrease of D(CE) for trivalent ions in trivalent neighbors goes parallel with a corresponding increase in ionic volume. However, comparison with Yb and Eu shows that ΔV/V alone will not account for this change. In addition, the difference in band structure, i.e. number of 5d electrons, and crystal structures (Gd and Dy are hexagonal, Sm is rhombic) will play an important role. Theoretical calculations are necessary for a more quantitive comparison.

To study the C.E. contributions to the magnetic field as proposed by Hüfner [1], the magnetic splitting

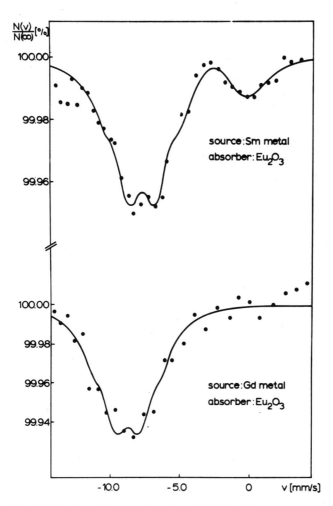

Fig. 2. Relative Transmission $N(v)/N(\infty)$ versus Doppler shift v for the 103 keV γ-rays of $^{153}$Eu.

TABLE III.- Contributions to the hyperfine fields $H_{eff}$ (in kOe) at nuclei in Eu, Gd, and Sm metal.

| | Eu [1] | Gd [2] | Eu in Gd | Eu in Sm |
|---|---|---|---|---|
| Core polarization | −340 ± 20 | −340 ± 20 | −340 ± 20 | −340 ± 20 |
| Conduction-electron polarization by own 4f electrons | +190 ± 20 | +250 ± 30 | +250 ± 50 | (+250 ± 50) |
| Neighbor effects: Conduction-electron polarization + overlap + covalency | −115 ± 20 | −260 ± 25 | −260 ± 60 | (−260 ± 60) |
| Total $H_{eff}$ measured | −265 ± 5 | −350 ± 13 | −290 ± 35 | −310 ± 25 |

of the 103 keV MB-transition in $^{153}$Eu in Gd- and Sm-
metals was measured. In Fig. 2 the MB spectra are shown.
The results are compiled in Table II together with
Hüfner's value of Eu in Eu, and the value of Gd in Gd
[2]. The fields, when decomposed into terms as in [1],
are consistent with a positive contribution due to C.E.
polarization by the metal's own 4f-electrons and a
negative term due to effects of neighbors. The values
are given in Table III.

REFERENCES

1. HÜFNER, S., *Phys. Rev. Lett. 19*, 1034 (1967).

2. ZMORA, H., BLAU, M. and OFER, S., *Phys. Lett. 28A*, 668 (1969).

# HYPERFINE INTERACTION IN d-SHELL ELEMENTS*

C.E. JOHNSON

*Oliver Lodge Laboratory, University of Liverpool, England*

ABSTRACT

The sources of hyperfine interaction in d-shell elements in solids are discussed and Mössbauer Effect measurements for separating the contributions from contact and non-contact interactions are described. The non-contact fields are a valuable source of information on the electronic states and environment of atoms in non-metallic solids. In metals the situation is more complicated, and experimental evidence on the role of the conduction electrons is reviewed.

Magnetism (and hence magnetic hyperfine interaction) occurs in solids when an atom has a partially filled inner shell of electrons, since the spins of the outermost electrons are generally paired off in bonding orbitals. The incomplete shell may contain either d or f electrons. Although the magnetic interactions between the electrons and the nuclei are basically the same for all electrons, their relative orders of magnitude are different for the two shells. The origin of this difference lies in the fact that the d-electrons are less deep inside the atoms and hence they are very strongly influenced by the electrostatic field of the neighbouring

---

* Invited paper.

atoms, which quenches the orbital angular momentum. By contrast f-electron ions have large orbital angular momenta. It is logical, therefore, to consider d- and f-shell elements separately, and in this paper we shall describe hyperfine structure in d-shell elements. In particular we shall examine the dependence of the hyperfine interaction on the environment in order to see how it may be used to study the state of the magnetic ion and the local structure close to it.

The theory of magnetic hyperfine interactions in solids has been extensively discussed in the literature [1]. There are several contributions, and they may be considered to be of two fundamentally distinct kinds - "contact" (e.g. core polarization) and non-"contact" (e.g. orbital and dipolar). The contact interaction, originally proposed by Fermi, is an interaction between s-electrons and the nuclei. Since the exchange interaction between d-electrons and the core s-electrons depends upon the relative orientation of their spins, the s-electrons become slightly polarized, an effect which has recently been observed in X-ray photo-electron spectra [2]. Because they have a finite density at the nucleus, the resulting hyperfine interaction is large. This interaction is isotropic, so that for any direction of magnetization it may be represented by an effective magnetic field at the nucleus (though it is not a real field - it depends upon what is sensing it) which may be written

$$H_c = \frac{2}{3} \mu_B \chi <g_s J> \qquad (1)$$

where $\mu_B$ is the Bohr magneton, $g_s J$ is the number of unpaired d-electrons and the parameter $\chi$ is a measure of the core polarization

$$\chi = \frac{4\pi}{g_s J} \sum_{n,s} [|\psi_{ns\uparrow}(0)|^2 - |\psi_{ns\downarrow}(0)|^2] \qquad (2)$$

where $\psi_{ns\uparrow}(0)$ is the wave function of the ns-electrons with spin up and the sum is over all n closed shells of s-electrons. When $\chi$ is given in atomic units, $H_c$ is

TABLE 1. Core Polarization Hyperfine Interaction

| Electron shell | $\chi$ (a.u.) | $H_c/g_s J$ $(kG/\mu_B)$ |
|---|---|---|
| 3d | -3.0 | -125 |
| 4d | -8.5 | -360 |
| 5d | -13.0 | -550 |

equal to $4.21 \times 10^4 g_s J \chi$ gauss. A useful quantity is $H_c/g_s J$, the contact field per unpaired electron. Table 1 gives some order-of-magnitude values of $\chi$ and $H_c/g_s J$ for d-shell elements. An alternative way of writing the contact field is in terms of $<r^{-3}>$, the average value of $1/r^3$ for the d-electrons, so that it may be compared with the non-contact terms. It is then

$$H_c = -2\mu_B <r^{-3}> \kappa <S> \qquad (3)$$

where $\kappa$ now is the parameter measuring the core polarization. When the orbital angular momentum is completely quenched the number of unpaired electrons becomes simply 2S, where S is the spin on the atoms. A great deal of attention has been given by theorists to the computation of $\chi$ for d-shell ions using spin-polarized Hartree-Fock (SPHF) calculations of the atomic wave functions. It turns out that $\chi$ is negative and almost constant for a particular d-shell, but increases with the principal quantum number of the shell, which is in good agreement with the experimental observations.

Experimentally, the hyperfine field for an S-state atom in different environments is found to vary, and this may be ascribed to covalency, i.e., to the transfer of d-electrons from the metal atom on to the ligand atoms. Table 2 gives some values for $Fe^{3+}$ compounds, where the hyperfine field is almost entirely contact in origin. If the electron wave function is written as a molecular orbital $\psi = N(\psi_d + \alpha\psi_{lig})$ it is seen that the free ion value of the d-electron density $|\psi_d|^2$ will be reduced by a factor $N^2$, and hence $\chi$ and $H_c$ will be reduced by

TABLE 2. Variation of $H_n$ with Covalency for Octahedrally Co-ordinated $Fe^{3+}$

| Ligands | $H_n$(kG) | Reference |
|---|---|---|
| $F^-$ | 622 | a |
| $H_2O$ | 584 | b |
| $O^{2-}$ | 540 | c |
| $Cl^-$ | 487 | d |
| $S^{2-}$ | 427 | e |

a. D.N.E. Buchanan and G.K. Wertheim, *Bull. Am. Phys. Soc. II7*, 227 (1962).
b. L.E. Campbell and S. de Benedetti, *Phys. Letters 20*, 102 (1966).
c. T. Nakamara and S. Shimizu, *Bull. Inst. Chem. Res. Kyoto Univ. 42*, 299 (1964).
d. C.W. Kocker, *Phys. Letters 24*, 93 (1967).
e. R. Richards, C.E. Johnson and H.A.O. Hill, *J. Chem. Phys. 48*, 5231 (1968).

covalency. For the highly ionic ligand $F^-$, $H_c/2S = 124$ kG/$\mu_B$, while for covalent $S^{2-}$ ligands $H_c/2S$ is reduced to 85 kG/$\mu_B$.

The non-contact hyperfine interactions are sensitive to and characteristic of the environment of the ion, and these therefore contain the information about the material being studied which is of interest to the solid state physicist, the metallurgist, the chemist or the biochemist, depending upon the nature of the sample. In general these terms are anisotropic so they are not correctly represented by magnetic fields but are tensor quantities. However, it is convenient to express them in units of magnetic field provided the direction of magnetization for which they are measured is specified. There are two contributions - one from the orbital moment, and one from the dipolar field of the electron spins. The component of the orbital field in the direction i is

HYPERFINE INTERACTION IN d-SHELL ELEMENTS 807

$$H_{Li} = 2\mu_B \langle r^{-3}\rangle \langle L_i\rangle \qquad (4)$$

This may also be written

$$H_{Li} = 2\mu_B \langle r^{-3}\rangle g_{Li}\langle J_i\rangle \qquad (5)$$

or, if the orbital angular momentum is almost completely quenched,

$$H_{Li} = 2\mu_B \langle r^{-3}\rangle (g_i - 2)\langle S_i\rangle \qquad (6)$$

The spin dipolar hyperfine interaction gives a field

$$H_d = 2\mu_B \langle r^{-3}\rangle \langle S - 3\hat{r}(S\cdot\hat{r})\rangle \qquad (7)$$

which has components

$$H_{di} = 2\mu_B \langle r^{-3}\rangle(-3\xi)\Big\{\langle L_i^2 - \tfrac{1}{3}L(L+1)\rangle\langle S_i\rangle$$
$$+ \tfrac{1}{2}\langle L_iL_j+L_jL_i\rangle\langle S_j\rangle + \tfrac{1}{2}\langle L_iL_k+L_kL_i\rangle\langle S_k\rangle\Big\} \qquad (8)$$

where $\xi$ is a reduced matrix element which depends upon the state of the ion. In these expressions $\langle r^{-3}\rangle$ is the effective value of $1/r^3$ averaged over the d-electrons; this value is $N^2(1 - R)$ times the free ion value, where $N^2$ represents a reduction in the fields arising from covalency and $1 - R$ takes account of the shielding of the nucleus from the d-electrons by the core electrons. In simple cases the dipolar hyperfine field may be related to the electric field gradient at the nuclei, since it arises from the anisotropy of electronic magnetic moment, whereas the electric field gradient arises from anisotropy of electronic charge [3].

Ions of f-electron elements have large orbital angular momenta (except when the ion is in an S-state, e.g. for $Gd^{3+}$) so that the non-contact fields dominate. For d-electron elements, however, contact and non-contact contributions are generally comparable with each other, so in order to interpret hyperfine field data one must be able to separate the two kinds of field. This may be attempted by assuming a value for $H_c$, either from a theoretical calculation, or by taking the measured value for an S-state ion in the same period (e.g. $Fe^{3+}$ for the 3d elements) since the non-contact terms are zero for an S-state. Alternatively, it is possible in certain cases to separate the contact from the non-contact fields experimentally either by (i) direct measurements of the anisotropy of the hyperfine interaction or (ii) measurements of anomalous hyperfine interaction.

Measurements of hyperfine interactions in solids may be made by several techniques, each with its own particular advantage or disadvantage. In general the resonance methods (EPR, NMR and the Mössbauer Effect) are powerful in measuring local differences of environment, and for detecting anisotropic effects. Other methods (nuclear orientation, perturbed angular correlations, nuclear specific heats) are valuable in being applicable to many elements not accessible to the resonance methods, but they have the limitation of measuring an average field only over the whole specimen, and cannot resolve different fields due to ions with different environments.

The Mössbauer Effect, although it is limited in practice to relatively few elements, has been used in two kinds of detailed study in which the contact field is separated from the non-contact field. They are (i) measurements of the anisotropy of the hyperfine interaction of $^{57}Fe$ in single crystals of iron salts and (ii) measurements of the anomalous hyperfine interaction of $^{193}Ir$ in iridium compounds.

ANISOTROPY OF HYPERFINE INTERACTION IN IRON SALTS

This kind of measurement is of course very common in EPR spectroscopy. A single crystal of a paramagnetic salt is magnetized in an external field applied along different directions, and the hyperfine splitting of its spectrum is observed and is found to depend upon the relative orientation of the field and the crystal axes. Analogous measurements may be made on Mössbauer spectra. In this case, however, hyperfine interaction may be observed in an applied magnetic field even if the electron spin relaxation times are short, and the effective magnetic field $H_i^{eff}$ at the nucleus is then measured, and is proportional to the magnetization. The hyperfine field ($H_{ni} = H_c + H_{Li} + H_{di}$) is the value of this field extrapolated to magnetic saturation, and is obtained from

$$\frac{H_{ni}^{eff}}{H_{ni}} = \frac{\chi_i H}{N\mu_i} \qquad (9)$$

where $\chi_i$ and $N\mu_i$ are the components of the susceptibility and the saturation magnetic moment per mole. Experimentally, it is simplest when the crystal has only one kind of ion in the unit cell, so that the crystal axes for all the magnetic ions are all parallel and the field makes unique angles with them. By way of examples we shall summarize results on $^{57}$Fe in a high spin $Fe^{2+}$ salt (ferrous fluosilicate) and in a low spin $Fe^{3+}$ salt (potassium ferricyanide).

*(i) High Spin $Fe^{2+}$*

Here the iron has a $(3d)^6$ $^5D$ configuration and the crystal field produces a large splitting of the orbital levels and spin-orbit coupling splits the spin states. Ferrous fluosilicate $FeSiF_6 \cdot 6H_2O$ is one of the simplest of these salts, having two ferrous ions per unit cell whose axes are identical. The symmetry is trigonal and the orbital ground state is $|L_z = 0\rangle$, where z is the trigonal axis. The orbital moment is strongly quenched, with $g_z = 2.00$ and $g_x = g_y = 2.14$. Using (2), (6) and (8) and $\xi = -1/42$,

$$H_i = 4\mu_B \langle r^{-3} \rangle [-\kappa + (g_i - 2) - \frac{1}{14}\langle L_i^2 - 2\rangle] \qquad (10)$$

The component of the dipolar field in the x direction is most easily found using the relation $\langle L_x^2 - 2 \rangle = (1/2)\langle L_z^2 - 2 \rangle$. By measuring the effective fields from the Mössbauer spectra when a field is applied along the x and z directions, the hyperfine field tensor was deduced [4]. The results show that $H_c = -422$kG, i.e. that $H_c/2S = -105$kg/$\mu_B$. This is less than the calculated value given in Table 1, but is close to the

experimental value $H_c$ = -550 kG found in ionic high spin
(S = 5/2) $Fe^{3+}$ salts, which give $H_c/2S$ = -110 kG/$\mu_B$.
The components of the non-contact fields were $H_{Lz}$ = 0,
$H_{Lx}$ = +107 kG, $H_{dz}$ = -128 kG and $H_{dx}$ = +64 kG.

*(ii) Low Spin $Fe^{2+}$*

When $Fe^{3+}$ is bonded to highly covalent ligands,
e.g. $-CN^-$, the ligand field is stronger than the electron-electron interaction, which tends to keep the
electron spins parallel (Hund's rule), and the five 3d
electrons are all in $t_{2g}$ orbitals. The resultant ion
has threefold orbital degeneracy ($L_{eff}$ = 1) and S = 1/2,
which are coupled by spin-orbit coupling to give a
ground state with J = 1/2. The strength of the spin-orbit coupling depends upon 1/2S and so is about four
times stronger than for the high spin $Fe^{2+}$ case, with
the result that the crystal field splittings in low spin
$Fe^{3+}$ are generally not large compared with the spin-orbit splitting, whereas the opposite is true for high
spin $Fe^{2+}$. In contrast to high spin ions the orbital
angular momentum is not quenched. Using (2), (5) and
(8) and $\xi$ = 2/21,

$$H_i = \mu_B <r^{-3}> [-\frac{g_s}{2}\kappa + g L i + \frac{2}{7}<L_i^2 - 2>] \qquad (11)$$

The best studied crystal is potassium ferricyanide,
$K_3Fe(CN)_6$. The symmetry is not too far removed from
cubic, but EPR data [5] had shown that the axes of the
g-tensor were rotated relative to the cubic axes, which
complicates matters by introducing off-diagonal elements
in the hfs (and other) tensors. The g-values are $g_x$ =
-0.91, $g_y$ = -2.34 and $g_z$ = -2.09, and the corresponding
values for the hyperfine tensors are -196, -277 and -247
kG [6]. Since the rotation of axes is about the z-axis
the interpretation of the component in this direction is
simplest. With the values $g_{sz}$ = -0.90, $g_{Lz}$ = -1.18, the
hyperfine field is resolved into components $H_c$ = +39,
$H_{Lz}$ = -218 and $H_{dz}$ = -68 kG. The contact term gives
$H_c/g_sJ$ = -86 kG/$\mu_B$, with the most ionic high spin values.

HYPERFINE INTERACTION IN d-SHELL ELEMENTS                811

HYPERFINE FIELDS IN METALS AND ALLOYS

Lastly, we describe applications of hyperfine field
measurements to the study of ferromagnetic metals and
alloys. A large number of measurements of this kind have
been made, using the $^{57}$Fe Mössbauer Effect and by other
methods (NMR, nuclear polarization, perturbed angular
correlations). For a long time there has been a great
need for a technique which would provide a local probe
of the electronic state of atoms in alloys, and hyperfine field measurements have helped to satisfy this
need. The Mössbauer Effect has been an especially valuable method, as the spectra of atoms with different
environments may be distinguished. We have seen that
it is possible to understand the magnetic hyperfine
interaction in crystals of non-conducting iron salts.
In these we know the orbital ground state from quadrupole splitting or susceptibility data and hence we can
calculate $H_L$ and $H_d$. In addition to the two examples we
have quoted, many high-spin $Fe^{2+}$ salts [7] and low spin
$Fe^{3+}$ haemoglobin complexes [8] have been studied, and
good agreement is found between theory and experiment.

In metals, however, the electronic state of the d-
shell atoms is not as simple as in non-conductors but
involves the band structure. It is generally not easy
to separate the contact from the non-contact contribution, but Perlow et al [9] have done this for iridium in
iron from measurements of the hyperfine anomaly in $^{193}$Ir.
It was found that the orbital field is $H_L = +$ 335 kG.
The positive sign is expected, since iridium in iron is
expected to lose about one electron to the conduction
band, i.e. it has about three holes in the 5d shell, and
for holes spin-orbit coupling aligns the orbit parallel
to the spin. This provides a welcome confirmation that
the theory of the orbital contribution to the hyperfine
interaction applies to the metallic state.

In pure iron, the hyperfine field is -330 kG. The
d-electron spin moment is 2.4 $\mu_B$ [10], so that the SPHF
calculated value of $H_c$ is -290 kG for a free atom. Since
simple covalency would reduce this value and since $H_L$ is

positive, an extra negative field must be found to explain the field.

In addition to the contributions already mentioned, in metals and alloys there is also a field at the nuclei, arising from the conducting electrons. This is an extra contact field and has two parts: (i) a field due to the polarization of the conduction electrons and (ii) a field due to admixing of s-electron into the d-band. A d-electron interacts with conduction electrons with the same spin direction via an interaction $V_{kd}$. This will cause the energy levels of the d and s states to repel each other, and will result in a mixing of their wave functions. If the spin up d-band is full and lies below the Fermi level, and the spin down d-band is empty and lies above the Fermi level, this interaction will increase the energy of the spin up s-electrons at the Fermi surface and will decrease the energy of the spin down s-electrons. Hence a negative conduction electron polarization is produced at the nucleus which has an oscillatory raidal distribution. This may outweigh the positive polarization of the s-electrons arising from the Pauli susceptibility. The effect of admixing s-electrons with the d-band is to produce an extra hyperfine field via the contact interaction and this will be positive in sign. Anderson and Clogston [11] have shown that the admixture and polarization effects will approximately cancel. Campbell [12] has analyzed data on the transferred hyperfine interaction at diamagnetic atoms (Ag, Cd, In, Sn, Sb, Te, etc.) dissolved in iron. He deduced that the conduction electron polarization of the iron matrix is negative and has explained the change in sign of the field between Sn and Sb from negative to positive. A negative conduction electron polarization has also been found from neutron diffraction data on iron [10].

When a small amount of an impurity is alloyed substitutionally with iron the $^{57}$Fe Mössbauer spectrum shows broader lines and the average hyperfine field changes compared with pure iron [13]. When the impurity is a transition metal to the right of iron in the periodic table (e.g. Co, Ni, Pt ), the broadening is relatively small, so that the lines do not show any resolved structure and the hyperfine field initially increases. For

HYPERFINE INTERACTION IN d-SHELL ELEMENTS 813

other impurities the line broadening is greater and the mean hyperfine field decreases. For all the iron alloys [13] the variation of the mean hyperfine with composition is qualitatively similar to that of the average atomic moment, i.e. to the Slater-Pauling curve.

The line broadening arises from the local variation of magnetization caused by the different arrangements of neighbouring impurity atoms, and therefore offers the possibility of more detailed studies than are available from measurements of average hyperfine fields. For alloys containing a few per cent of impurity (except those with transition metals with more d-electrons than iron), the spectrum due to iron atoms with one impurity nearest neighbour can usually be resolved from those which are distant from an impurity [14]. The distant neighbours have a field which is close (usually about 1% greater) to that of pure iron. The first neighbour atoms have a field which is less than the unperturbed value. For non-transition impurities (e.g. Al, Si, Sn) and for 3d elements lying to the left of iron in the periodic table this reduction is about 7% of the field. For 4d elements (e.g. Mo) the reduction is about 10% [15], while for 5d elements (e.g. W, Re) it is about 15% [16]. Between the first and distant neighbour lines it is sometimes possible to observe lines due to second and even third neighbours. The range of the disturbance produced by impurity atoms in iron is relatively small, however. For example, in Fe-Si alloys only the first neighbours have a hyperfine field which is detectably different from the unperturbed value - the second neighbours have a field which is the same as that of distant neighbours [17]. Thus the conduction electrons strongly screen the impurity magnetic moments at distances greater than first neighbours.

This screening is reflected in the smallness of the anisotropy of the hyperfine interaction which is observed in the metallic state. In hexagonal cobalt, for example, the difference between the hyperfine field of $^{57}$Fe when the magnetization is parallel and perpendicular to the c-axis is less than 0.5 kG in 310 kG [18]. In iron alloys where the cubic symmetry at the iron sites is destroyed by the presence of a neighbouring impurity, single crystal measurements have detected an anisotropy of only about 2 or 3 kG [17], i.e. less than 1% of the total hyperfine field.

The decrease in hyperfine field observed at iron neighbours in transition metal and non-transition metal alloys is much larger than would be expected from the change in charge density measured by the isomer shift. In alloys with 3d elements lying to the left of iron in the periodic table the iron neighbours to an impurity have a very small and negative isomer shift relative to more distant iron atoms. In non-transition metal alloys the shift is small but positive. However, the sign of the change in hyperfine field at these atoms is the same (i.e. a decrease) for both these kinds of alloy. Clearly, the interpretation of hyperfine field data in alloys is not simple and is less direct than that of isomer shift measurements, presumably because it depends upon the details of the shape of the d-band and of its splitting by the exchange field.

REFERENCES

1. FREEMAN, A.J., *Hyperfine Structure and Nuclear Radiations*, ed. Matthias and Shirley, North-Holland, 427 (1968).

2. FADLEY, C.S., SHIRLEY, D.A., FREEMAN, A.J., BAGUS, P.S. and MALLOW, J.V., *Phys. Rev. Lett.* 23, 1397 (1969).

3. MARSHALL, W. and JOHNSON, C.E., *J. Phys. Rad.* 23, 733 (1962).

4. JOHNSON, C.E., *Hyperfine Structure and Nuclear Radiations*, ed. Matthias and Shirley, North-Holland, 226 (1968); *Proc. Phys. Soc.* 92, 748 (1967).

5. BAKER, J.M., BLEANEY, B. and BOWERS, K.D., *Proc. Phys. Soc.* B69, 1205 (1966).

6. OOSTERHUIS, W.T. and LANG, G., *Phys. Rev. 178*, 439 (1968).

7. e.g. JOHNSON, C.E., *Symposium of the Faraday Society 1*, 7 (1967).

8. LANG, G. and MARSHALL, W., *Proc. Phys. Soc. 87*, 3 (1966).

9. PERLOW, G.J., HENNING, W., OLSON, D. and GOODMAN, G.L., *Phys. Rev. Lett. 23*, 680 (1969).

10. SHULL, C. and YAMADA, Y., *J. Phys. Soc. Japan 17*, Supp BIII, 1 (1962).

11. ANDERSON, P.W. and CLOGSTON, A.M., *Bull. Am. Phys. Soc. 6*, 124 (1961).

12. CAMPBELL, I.A., *J. Phys. C. 2*, 1338 (1969).

13. JOHNSON, C.E., RIDOUT, M.S. and CRANSHAW, T.E., *Proc. Phys. Soc. 81*, 1079 (1963).

14. WERTHEIM, G.K., JACCARINO, V., WERHICK, J.H. and BUCHANAN, D.N.E., *Phys. Rev. Lett. 12*, 24 (1964).

15. MARCUS, H.L. and SCHWARTZ, L.H., *Phys. Rev. 162*, 259 (1967).

16. BERNAS, H. and CAMPBELL, I.A., *Solid State Comm. 4*, 577 (1966).

17. CRANSHAW, T.E., JOHNSON, C.E., RIDOUT, M.S. and MURRAY, G., *Phys. Lett. 21*, 481 (1966).

18. PERLOW, G.J., JOHNSON, C.E. and MARSHALL, W., *Phys. Rev. 140*, A875 (1965).

# MÖSSBAUER EFFECT IN MIXED Er IRON GARNETS AT 4.2°K

U. ATZMONY, F. T. PARKER and J. C. WALKER

*The Johns Hopkins University, Baltimore, Maryland*

E. L. LOH

*Towson State College, Baltimore, Maryland*

The Mössbauer effect of the $2^+ \to 0^+$ transition in $Er^{166}$ has been remeasured in ErIG and also measured in the mixed garnets $Er_{1/3}Sm_{2/3}IG$ and $Er_{1/3}Ho_{2/3}IG$. From these measurements and from previous measurements with mixed europium iron garnets [1] we conclude that it is possible to interpret the ErIG spectrum in terms of a [100] direction of easy magnetization if one assumes some canting of of the iron moments with respect to this direction.

In ErIG, at 4.2°K, Hüfner et al. [2] interpreted the Mössbauer spectrum as the result of *two* magnetically inequivalent sites of relative populations 3:3 with a magnetic field ratio about 1.4. This suggests that [111] is the direction of easy magnetization. This result disagrees with optical and magnetization measurements which suggest that the direction of easy magnetization at 4.2°K is [100] [3]. Without canting, this would involve two inequivalent sites with relative populations 4:2. Hüfner et al. could not fit their spectra satisfactorily with this ratio.

The ErIG spectrum (Fig. 1d) obtained in this laboratory was quite similar to that of Hüfner et al. We have fit the ErIg spectrum assuming four sites with relative

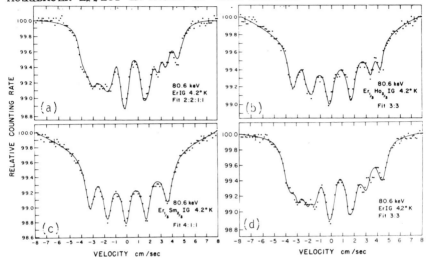

Fig. 1.  a) ErIG spectrum fit with 2:2:1:1 site occupation appropriate to a [100] magnetization with canting of the iron moments.
b) $Er_{1/3}Ho_{2/3}IG$ spectrum fit assuming a [111] magnetization direction.
c) $Er_{1/3}Sm_{2/3}IG$ spectrum fit assuming a [110] direction of magnetization.
d) ErIG spectrum fit assuming a [111] direction of magnetization.

occupation 2:2:1:1 which is a possible result of canting of the iron moments with respect to the [100] direction (Fig. 1a). We have fit the $Er_{1/3}Ho_{2/3}IG$ (Fig. 1b) assuming a [111] direction of easy magnetization, and Er $Er_{1/3}Sm_{2/3}IG$ (Fig. 1c) assuming a [110] direction. There is, at present, insufficient data to establish the canting angle, but it does appear that this model can reconcile the magnetization, optical and Mössbauer effect measurements.

REFERENCES

1. ATZMONY, U. et al., *Phys. Rev. 179*, 514 (1967).
2. HÜFNER, S. et al., *Proc. Int. Conf. on Magn.*, Nottingham (1964), p. 672.
3. ORLICH, E. and HÜFNER, S., *J. Appl. Phys. 40*, 1503 (1969).

# MÖSSBAUER EFFECT MEASUREMENTS ON DyP, DyAs, DySb

G. VÉCSEY and W. DEY

*Laboratory for High Energy Physics, Swiss Federal Institute of Technology, Zurich, Switzerland*

The absorption spectra of the magnetically ordering trivalent Dy-compounds were measured at low temperatures.

The 25.6 keV γ-rays of a room temperature enriched $^{160}Gd_2O_3$ powder source were detected by a lead loaded plastic scintillator at a counting rate of about $10^6$/sec and were registered in the multiscaling mode. The temperature of the absorber was varied between 3.3 and 16°K. Magnetic ordering was observed for the ferrimagnetic DyP and DyAs at 8 and 7.5°K respectively, and for the antiferromagnetic DySb at 12°K. Some typical spectra are plotted in Fig. 1.

Assuming a ground state doublet for the $Dy^{3+}$ ions and neglecting relaxation effects, the 14 line Hfs taken at the lowest temperatures were fitted and have given the values listed in Table 1.

TABLE 1. - Hyperfine Interaction in Mc/s

|  | DyP (3.3°K) | DyAs (4.2°K) | DySb (4.6°K) |
|---|---|---|---|
| $g_0\beta_N H_{eff}$ | 764 ± 20 | 753 ± 20 | 797 ± 20 |
| $\frac{1}{4} Q_0 eV_{zz}$ | 640 ± 30 | 640 ± 30 | 670 ± 30 |

# MÖSSBAUER EFFECT MEASUREMENTS 819

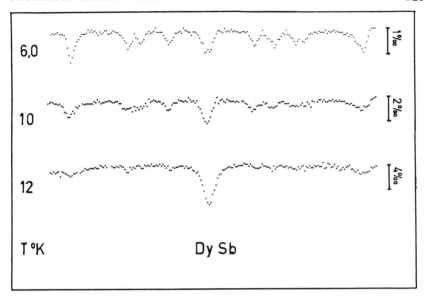

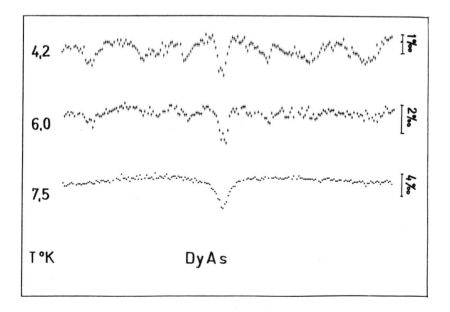

Fig. 1. Hyperfine Spectra of DyAs and DySb

The observed ordering temperatures are different from those obtained by bulk high field magnetization measurements [1]. In the case of DyAs and DySb, however, they agree well with the values at which spontaneous magnetostriction was observed [2].

REFERENCES

1. BUSCH, G. *J. Appl. Phys. 38*, 1386 (1967).

2. LÉVY, F. *Physik kondens. Materie 10*, 85 (1969).

# INTERFERENCE BETWEEN HYPERFINE-STRUCTURE STATES IN MÖSSBAUER SCATTERING

K. GABATHULER* and H.J. LEISI

*Laboratory for High-Energy Physics,
Swiss Federal Institute of Technology,
Zurich, Switzerland*

The velocity spectra of Mössbauer scattering experiments for powder absorbers exhibiting hyperfine splitting have usually been analyzed in terms of a superposition of pure Lorentzian-shaped lines [1]. The purpose of this note is to show that there are important deviations from the Lorentzian shape, apart from the effects due to the interference between Mössbauer and Rayleigh scattering [2]. These deviations result from the interference between the hyperfine structure levels of the excited nuclear state.

In the case of unpolarized incident radiation from an unsplit source (natural width $\Gamma$) and an equidistant magnetic hyperfine splitting in the absorber (ferromagnetic powder), the differential cross-section for Mössbauer scattering can be written as [3]

---

*Present address: CERN, Synchro-cyclotron Machine Division, 1211 Geneva 23, Switzerland.

$$\frac{d\sigma(v)}{d\Omega} =$$

$$\sigma' \sum_{k,m_i} \left\{ \sum_m B_k(m_i;m,m) \frac{\Gamma^2}{[E_0(1+v/c)-E_{mm_i}]^2+\Gamma^2} + \sum_{\substack{m,m' \\ m'>m}} B_k(m_i;m,m') \times \right.$$

$$\left. \frac{[E_0(1+v/c)-E_{mm_i}][E_0(1+v/c)-E_{m'm_i}][\Delta^2_{mm'}+2\Gamma^2]\Gamma^2 + 2\Gamma^6}{[(E_0(1+v/c)-E_{mm_i})^2+\Gamma^2][(E_0(1+v/c)-E_{m'm_i})^2+\Gamma^2][\Delta^2_{mm'}+\Gamma^2]} \right\} \times$$

$$P_k(\cos\theta). \qquad (1)$$

Here $E_0$ is the nuclear energy difference in the source,

$$E_{mm_i} = E_0 - g_e\mu_N Hm + g_g\mu_N Hm_i \qquad (2)$$

is the energy difference between the excited state $|Im\rangle$ and the ground state $|I_i m_i\rangle$,

$$\Delta_{mm'} = g_e\mu_N H(m' - m) \qquad (3)$$

is the energy splitting of two levels of the excited state. The interference amplitude $B_k(m_i;m,m')$ is defined by

$$B_k(m_i;m,m') = (1+\delta^2)^{-2} \left\{ G^{(mm')}_{km_i}(II_iLL) + \delta\left[G^{(mm')}_{km_i}(II_iLL') + G^{(mm')}_{km_i}(II_iL'L)\right] + \delta^2 G^{(mm')}_{km_i}(II_iL'L') \right\} \times$$

$$[F_k(II_iLL) - 2\delta F_k(II_iLL') + \delta^2 F_k(II_iL'L')], \quad (4)$$

where

INTERFERENCE BETWEEN HFS STATES                 823

$$G_{km_i}^{(mm')}(II_iLL') =$$

$$(-1)^{m+m'-I-m_i+1}[(2k+1)(2L+1)(2L'+1)(2I+1)]^{\frac{1}{2}} \times$$

$$\begin{Bmatrix} L & L' & k \\ 1 & -1 & 0 \end{Bmatrix} \begin{pmatrix} I_i & L & I \\ m_i & m-m_i & -m \end{pmatrix} \begin{pmatrix} I_i & L' & I \\ m_i & m'-m_i & -m' \end{pmatrix} \times$$

$$\begin{pmatrix} L & L' & k \\ m_i-m & m'-m_i & m-m' \end{pmatrix} \begin{pmatrix} I & I & k \\ m' & -m & m-m' \end{pmatrix}, \quad (5)$$

$F_k(II_iLL')$ are the usual angular correlation coefficients [4] and the mixing ratio $\delta$ is defined by the reduced matrix elements for emission

$$\delta = \frac{<I_i||L'||I>}{<I_i||L||I>} ; \quad L' = L + 1. \quad (6)$$

The constant $\sigma'$ is given by

$$\sigma' = \left(\frac{\hbar c}{E_0}\right)^2 \frac{f(2I+1)}{2(2I_i+1)(1+\alpha)^2}, \quad (7)$$

The other symbols have their usual meaning.

The second term in Eq. (1), which represents the interference part, not only causes deviations from the Lorentzian line shape, but also affects the scattered intensity at a given scattering angle $\theta$. The scattered intensity as a function of the velocity v is shown in Fig. 1 for a particular situation.

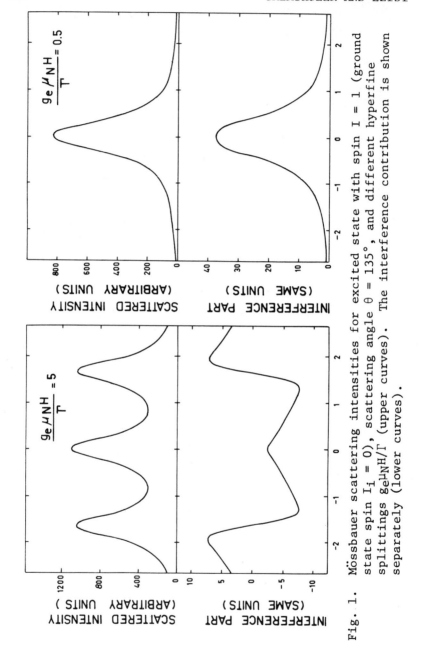

Fig. 1. Mössbauer scattering intensities for excited state with spin $I = 1$ (ground state spin $I_i = 0$), scattering angle $\theta = 135°$, and different hyperfine splittings $g_e\mu_N H/\Gamma$ (upper curves). The interference contribution is shown separately (lower curves).

REFERENCES

1. ATAC, M., DEBRUNNER, P. and FRAUENFELDER, H.
   *Phys. Lett. 21*, 699 (1966).
   TUMOLILLO, T.A. *Nuclear Phys. A143*, 78 (1970).

2. HANNON, J.P. and TRAMMEL, G.T. *Phys. Rev. 186*, 306 (1969).

3. GABATHULER, K., Diplomarbeit, Swiss Federal Institute of Technology, Zurich (1970) (unpublished).

4. FERENTZ, M. and ROSENZWEIG, N., in *Alpha-, Beta- and Gamma-Ray Spectroscopy* (edited by K. Siegbahn) (North-Holland Publ. Co., Amsterdam, 1965), Appendix 8.

# MÖSSBAUER EFFECT STUDIES OF RELAXATION PHENOMENA IN $NH_4Fe(SO_4)_2 \cdot 12H_2O$

S. MØRUP and N. THRANE

*Laboratory of Applied Physics II, Technical University of Denmark, Lyngby, Denmark.*

The relaxation broadened Mössbauer line of polycrystalline $NH_4Fe(SO_4)_2 \cdot 12H_2O$ (Fe alum) has previously been found to become narrower when small external magnetic fields are applied [1,2]. The present paper reports an investigation of this phenomenon in the temperature range 90-300°K and with external fields up to 5 kG applied perpendicular to the γ-ray direction.

In the whole temperature range a decrease of the line width (FWHM) takes place in fields up to 1 kG, resulting in an increase of the line amplitude. Representative data obtained at 200°K are shown in Fig. 1. At temperatures below about 150°K and external fields larger than 1 kG the line appears to consist of a broad and a narrow component.

In Fe alum diluted with Al the crystal field splitting of the $Fe^{3+}$ S-state ions is small, about 0.12°K [3]. It is reasonable to assume a splitting of the same order in the pure compound. The fluctuating magnetic field $H_i$ on the Fe ion arising from the surrounding Fe ions is in average about 450 G [4]. Hence this magnetic interaction is comparable to the crystal field interaction. $H_i$ will decouple the Fe nucleus from the ionic spin, and the hyperfine interaction can therefore be described by means of an effective magnetic field, $H_h$.

In the following discussion we shall neglect the

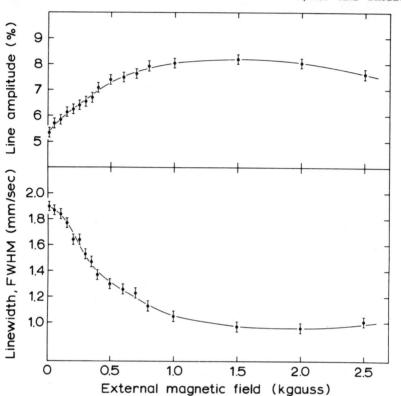

Fig. 1. Line amplitude and line width (FWHM) of the Mössbauer absorption line of $NH_4Fe(SO_4)_2 \cdot 12H_2O$ at 200°K as a function of the external magnetic field applied perpendicular to the γ-ray direction.

crystal field interaction as it has only a little influence on the results.

When the external field $H_0$ is large compared to $H_i$, the electronic spins will precess around the z-direction defined by $H_0$. $H_h$ will therefore be parallel to the z-direction, but because of electronic relaxation it will fluctuate both in sign and magnitude. In this case the Mössbauer spectrum will consist of a narrow component arising from the $(\pm \frac{1}{2} \to \mp \frac{1}{2})$ nuclear transition and broader components arising from the $(\pm \frac{1}{2} \to \pm \frac{1}{2})$ and $(\pm \frac{3}{2} \to \pm \frac{1}{2})$ nuclear transitions, assuming a relaxation time of $10^{-9} - 10^{-11}$ sec [5,6].

We have estimated that for relaxation times of about $10^{-9}$ sec the line width (FWHM) is mainly determined by the narrow $(\pm \frac{1}{2} \to \mp \frac{1}{2})$ component, the other components being very broad and therefore of small amplitude.

When $H_0$ is comparable to or smaller than $H_i$, $H_h$ will in the average be parallel to the z-direction. However, at any moment the direction of $H_h$ will be determined by the vector sum of $H_i$ and $H_0$. Hence $H_h$ will have a component fluctuating perpendicular to the z-direction. In this case transitions between the nuclear sublevels can be induced resulting mainly in a broadening of the narrow $(\pm \frac{1}{2} \to \mp \frac{1}{2})$ component [5].

The observed narrowing of the Mössbauer line with application of external magnetic fields as demonstrated in Fig. 1 can therefore be explained on the basis of a hyperfine field fluctuating in a fixed direction.

The high field spectra, $H_0 = 5$ kG, have been fitted with computed spectra based on the theory by Wegener [6], which should be valid in this simple case. The fitting procedure leads to an estimate of the relaxation time involved. Examples are shown in Fig. 2. The integral intensity ratios of the $(\pm \frac{3}{2} \to \pm \frac{1}{2})$, $(\pm \frac{1}{2} \to \pm \frac{1}{2})$, and $(\pm \frac{1}{2} \to \mp \frac{1}{2})$ nuclear transitions were assumed to be 3:4:1 as $H_0$ was applied perpendicular to the γ-ray direction. This assumption has been justified by successful fitting of spectra obtained with $H_0 = 5$ kG applied parallel to the γ-ray direction using integral intensity ratios of 3:0:1 and the same relaxation time as in the perpendicular case at the same temperature. In these spectra the narrow component is very dominant.

The spectra shown in Fig. 2 demonstrate a temperature dependence of the relaxation time. This dependence must be due to spin-lattice relaxation as spin-spin re-

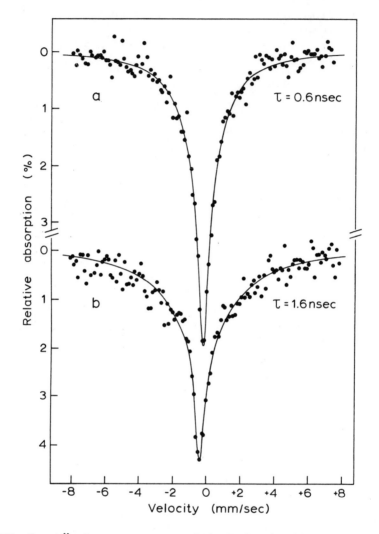

FIG. 2. Mössbauer spectra of $NH_4Fe(SO_4)_2 \cdot 12H_2O$ in external magnetic field 5 kG perpendicular to the γ-ray direction. a) T = 175°K b) T = 89°K. The solid curves represent theoretical spectra for the relaxation time indicated.

laxation is essentially independent of temperature. Measurements on Fe alum diluted with Al indicate a spin-lattice relaxation time of about $10^{-9}$ sec at room temperature [7,8]. The present measurements indicates a spin-lattice relaxation time of the same order in the pure compound at temperatures above 150°K.

REFERENCES

1.  HOUSLEY, R.M. and DE WAARD, H., *Phys. Lett. 21*, 90 (1966).

2.  HOUSLEY, R.M., *J. Appl. Phys. 38*, 1287 (1967).

3.  MEIJER, P.H.E., *Physica 17*, 899 (1951).

4.  VOGLER, I., DE VRIJER, R.W. and GORTER, C.J., *Physica 13*, 621 (1947).

5.  BLUME, M. and TJON, J.A., *Phys. Rev. 165*, 446 (1968).

6.  WEGENER, H., *Z. Physik 186*, 498 (1965).

7.  DE WAARD, H. and HOUSLEY, R.M., *Hyperfine Interactions* (Edited by A.J. Freeman), Academic Press, New York, 1967, 691.

8.  CAMPBELL, L.E. and DE BENEDETTI, S., *Phys. Rev. 167*, 556 (1968).

# THE INTERNAL MAGNETIC FIELD AND COVALENCY IN KNiF$_3$ AND THE NICKEL DIHALIDES

J.C. TRAVIS and J.J. SPIJKERMAN

*National Bureau of Standards, Washington, D.C.*

Previously reported internal fields of 99, 36, 75, 0, and 60 kG for NiF$_2$, NiCl$_2$, NiBr$_2$, NiI$_2$ and KNiF$_3$, all at 4.2°K, have been analyzed by use of a two parameter model for the relation of covalency to the internal field. The dipolar field contribution was neglected due to the high symmetry of these systems. In the limit of ionic bonding, values may be determined for $H_C'$, the Fermi contact term, and $H_L'$, the orbital term, from theoretical calculations and measured g values. Covalency may be included by the introduction of parameters $N_\sigma^2$ and $\alpha^2$ such that

(1) $\quad H_C = N_\sigma^2 H_C', \quad H_L = \alpha^2 H_L', \quad \alpha^2 = <r^{-3}>/<r^{-3}>_0$

$0.5 \leq N_\sigma^2 \leq 1.0, \quad 0.0 \leq \alpha^2 \leq 1.0.$

$N_\sigma^2$ is the square of a molecular orbital normalizing factor and is a measure of symmetry restricted covalency (SRC) [1]. The 'electron delocalization parameter, $\alpha^2$, can be influenced by both SRC and central field covalency (CFC) [1].

The availability of both a measured g value and a calculated $N_\sigma^2 = 0.937$ [2] for KNiF$_3$ permitted an evaluation of $\alpha^2 \simeq 0.8$ from our data. For the three halides with measured g values, plausibility arguments (based on the measured field values) may be advanced to define parameter ranges as follows:

(2)   $0.46 < \alpha^2(\text{NiX}_2) < 0.64$, $0.82 < N_\sigma^2(\text{NiF}_2) < 1.0$

$0.57 < N_\sigma^2(\text{NiCl}_2) < 0.75$, $0.57 < N_\sigma^2(\text{NiBr}_2) < 0.68$

The ranges of (2) may be narrowed to

(3)   $0.59 < \alpha^2(\text{NiX}_2) < 0.64$, $0.94 < N_\sigma^2(\text{NiF}_2) < 1.0$

$0.69 < N_\sigma^2(\text{NiCl}_2) < 0.75$, $0.65 < N_\sigma^2(\text{NiBr}_2) < 0.68$

if one is willing to impose the requirement that $N_\sigma^2$ be linear with Pauling electronegativity.

REFERENCES

1. JØRGENSEN, C.K., *Orbitals in Atoms and Molecules*, Academic Press, London, p. 59 (1962).

2. SUGANO, S. and SHULMAN, R.G., *Phys. Rev. 130*, 517 (1963).

# MÖSSBAUER INVESTIGATION OF JAHN-TELLER DISTORTION SPINELIC MANGANITES

G. FILOTI, A. GELBERG, V. GOMOLEA and M. ROSENBERG

*Inst. for Atomic Physics and Inst. of Physics, Bucharest, Romania*

Mössbauer spectra of the manganites $M^{2+}Mn^{3+}_{1.9}Fe^{3+}_{0.1}O_4$ were investigated (M = Cd, Mg, Zn, Co, Mn). We have studied the temperature dependence of the quadrupole splitting (QS) of the isomer shift and of peak areas in the 15-1300°C range comprising the transition from tetragonal to cubic symmetry.

The main results obtained were the following:

1. All the compounds enumerated above display two quadrupole doublets, which can be ascribed to a partial inversion of the spinels.

2. The QS of both doublets decrease with increasing temperature, indicating a gradual transition from tetragonal to cubic symmetry, due to the removal of the cooperative Jahn-Teller effect.

3. The area ratio of the two doublets is generally temperature dependent, obviously due to cation migration leading to a normalization of the spinel. In the case of $MgMn_{1.9}Fe_{0.1}O_4$ this process manifests itself as an anomaly of the QS vs. T curve, in agreement with electrical conductivity and X-ray diffraction measurements.

4. The QS is linearly dependent upon the distortion c/a - 1.

5. $Fe^{3+}$ ions are located at octahedral sites.

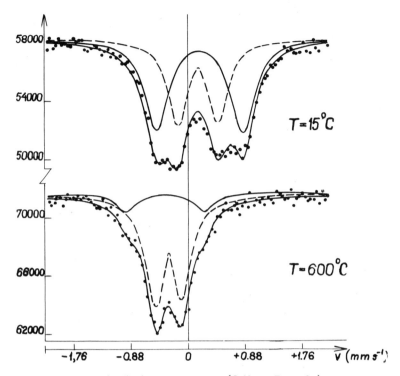

Fig. 1. Typical Mössbauer spectra (CoMn$_{1.9}$Fe$_{0.1}$O$_4$)

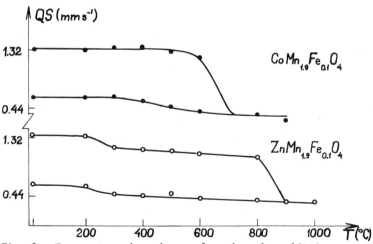

Fig. 2. Temperature dependence of quadrupole splitting.

# CRYSTAL FIELD EFFECTS FOR $Er^{3+}$ IN $HoAl_2$ STUDIED BY PAC

R. WÄPPLING, E. KARLSSON, G. CARLSSON and M.M. BAJAJ

*Institute of Physics, Uppsala, Sweden*

Magnetization data on the rare earth intermetallics with the cubic Laves' phase structure indicate that the angular momentum degeneracy of the rare earth ion is lifted by the crystal field. The effect on the hyperfine fields can be studied by the Mössbauer effect or by perturbed angular correlations (PAC). Relaxation effects and the magnitude of the recoil-free fraction make the Mössbauer method useful only at low temperatures, whereas PAC can be used over a wide temperature range. In the ferromagnetic region the exchange field splitting of the electronic sublevels might obscure the crystal field effects and it is therefore advantageous to work in the paramagnetic region. The crystal field effects for $Er^{3+}$ ions in $HoAl_2$ was accordingly investigated in the paramagnetic state.

In the presence of a magnetizing field $B_{ext}$, the degeneracy of the magnetic substates is lifted. The Boltzmann population of these levels gives a magnetization of the 4f-shell of the rare earth ion and a magnetic hyperfine field is created. When J is a good quantum number, the hyperfine field is given by

$$B_{hf} = 2\mu_B \langle r^{-3}\rangle \langle J||N||J\rangle \langle J_z\rangle$$

where other small contributions are neglected. $\langle J_z\rangle$ is the Boltzmann average of the different orientations of J at temperature T. The amplifying action of the electronic shells is usually described by the paramagnetic amplification factor $\beta$ through

# CRYSTAL FIELD EFFECTS FOR $Er^{3+}$

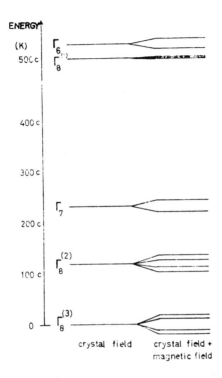

Fig. 1. Energy levels for the $^4I_{15/2}$ ground state of $Er^{3+}$

$$B_{tot} = B_{ext} + B_{hf} = \beta B_{ext}$$

In a crystal field the wavefunctions are not eigenfunctions of $J_z$, but can be expressed as linear combinations of these, provided that states of different J do not mix. This is the case for the ground term of $Er^{3+}$ ($^4I_{15/2}$).

The factor $\beta$ was calculated for different values of the strength of the crystal field given by the parameter c. (The total splitting of the ground term is given as 552.9 × cK). When the electric interaction is dominating we obtain the situation shown in Fig. 1, whereas in the other extreme we obtain energies close to the "free ion values." The quantity <J> can now be evaluated and

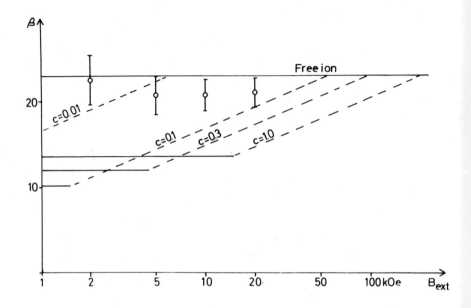

Fig. 2. Paramagnetic amplification factor β as function of magnetic field strength at 80°K.

the factor β was calculated as function of the strength of the external field and as function of temperature. The first alternative gives the highest sensitivity and the result of the calculation and the measurement at 80°K is shown in Fig. 2 and we can conclude that the cubic crystal field is indeed weak (< 10K).

# ISOMER SHIFTS AND HYPERFINE SPLITTING OF THE 145 keV MÖSSBAUER LINE OF $^{141}$Pr

W. KAPFHAMMER, W. MAURER, F.E. WAGNER and P. KIENLE

*Physik-Department, Technische Universität München, Munich, Germany*

The Mössbauer scattering of the 145 keV γ-rays of $^{141}$Pr was observed in a number of Pr compounds and in Pr metal. Large isomer shifts were found between Pr(III) and Pr(IV). From the hyperfine pattern of $PrO_2$ the magnetic moment of the 145 keV state, $\mu_{7/2} = (2.8 \pm 0.2)\mu_N$ was derived.

Many of the single proton transitions between the $2d_{5/2}$ and $1g_{7/2}$ nuclear states occuring in the $^{121}$Sb to $^{151}$Eu region are accessible to nuclear gamma resonance experiments. The 145 keV $7/2^+ \rightarrow 5/2^+$ transition in $^{141}$Pr, due to its high energy, requires the use of scattering techniques [1], for which it is well suited because of the simple decay scheme of the $^{141}$Ce source activity. The Mössbauer experiments reported here were performed under a scattering angle of 120° with both source and scatterer cooled in a liquid He cryostat. The scattered γ-rays were detected by a Ge(Li) counter capable of resolving the elastic peak from the Compton background. The $^{141}$Ce source activity ($T_{1/2}$ = 33 d) was produced by neutron bombardment of enriched $CeO_2$ containing 99.7% of the $^{140}$Ce isotope. After the irradiation the samples were annealed at 800°C for several hours. Only after this treatment would they emit the narrow line expected in cubic, diamagnetic $CeO_2$. The narrowest line, obtained for a 280 mg/cm$^2$ $PrF_3$ scatterer, had a full width at half maximum of 1.36 ± 0.02 mm/s or 1.4 times the natural width. Some of the Mössbauer spectra are reproduced in Fig. 1, a summary of the isomer shift

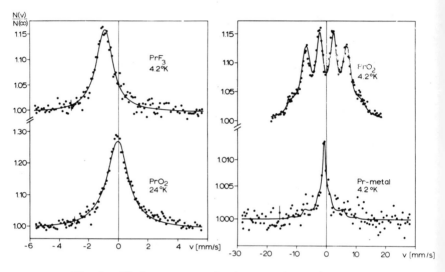

Fig. 1. Mössbauer scattering spectra of the 145 keV γ-rays of $^{141}$Pr measured with an annealed source of $^{141}$Ce in $CeO_2$ and various scatterers.

results is given in Fig. 2. There is virtually no shift between an annealed $CeO_2$ source and a $PrO_2$ scatterer, thus Pr is in its tetravalent state after the β⁻ decay in $CeO_2$. With respect to an unannealed source a shift of +0.56 ± 0.04 mm/s was observed for $PrO_2$, indicating the presence of trivalent praseodymium in the source due to radiation damage incurred during the neutron irradiation.

$PrO_2$ is antiferromagnetic below 14°K [2]. From its magnetic hyperfine pattern (Fig. 1) the ratio of g-factors $g_{7/2}/g_{5/2}$ = 0.48 ± 0.03 could be determined. With the known ground state moment one then obtains $\mu_{7/2}$ = (2.8 ± 0.2)$\mu_N$ for the magnetic moment of the 145 keV state, in good agreement with the result of Cook et al. [4]. The Mössbauer lines in $PrO_2$ at 4.2 and 1.8°K are broadened by about a factor of 3, the outer lines being more affected than the inner ones, which suggests some smearing out of the effective hyperfine field. In paramagnetic $PrO_2$ at 24°K the line width is only about 1.7 times the natural one. No significant differences were observed in the behaviour of $PrO_2$ samples prepared in different ways [5-7]. The average hyperfine field of 850 ± 30 kOe at 4.2°K is about a factor of 2.5 less than the field of 2180 kOe, which one estimates [8] for the $^2F_{5/2}$ ground state of the free $Pr^{4+}$ ion with the $Pr^{3+}$ value of 5.0 a.u. [9] for $<r^{-3}>$.

Unresolved hyperfine splittings, probably of magnetic origin, were also observed at 4.2°K in $PrFeO_3$, $PrC_2$, $CsPrF_5$, $Cs_2PrF_6$ and Pr metal. For the latter a complicated magnetic structure with half the spins paramagnetic and the other half antiferromagnetically ordered below 26°K has been proposed [10]. For the ordered spins the magnetic moment of the $Pr^{3+}$ ions is supposed to be sinusoidally modulated with a maximum value of $\mu_z = 0.95\mu_B$ [11]. A corresponding pattern was fit to the Mössbauer spectrum (Fig. 2) obtained with a scatterer of high purity Pr metal at 4.2°K. The maximum hyperfine field thus obtained is (860 ± 60) kOe. If conduction electron contributions to the hyperfine field are neglected this corresponds [8,9] to $\mu_z = (0.82 \pm 0.06)\mu_B$, in fair agreement with the specific heat result of $\mu_z = 0.95\mu_B$ [11]. From the concept of half the spins being paramagnetic the intensity of the unsplit line relative to the split pattern is expected to be unity. The experimental value of 0.6 ± 0.3 for this ratio is in fairly good agreement with expectation and shows that oxide impurities do not seriously contribute to the central line, which would lead to values larger than one.

The isomer shifts for trivalent Pr compounds (Fig. 2) are found near -0.85 mm/s in a fairly narrow region, as would be expected from the results for heavier rare earths like Eu [12,13]. Tetravalent Pr is known in compounds with oxygen and fluorine only. The isomer shift between the complex fluorides $CsPrF_5$ and $Cs_2PrF_6$ [14]

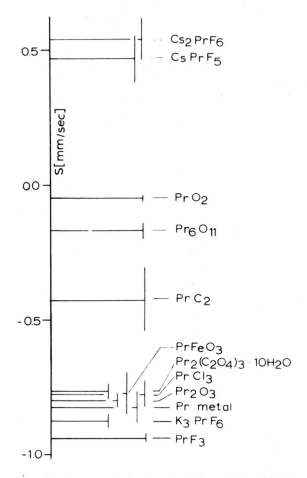

Fig. 2. Summary of the isomer shifts of the 145 keV γ-rays of $^{141}$Pr obtained with various scatterers at 4.2°K. The shifts are given with respect to an annealed source of $^{141}$Ce in $CeO_2$. For $Pr_6O_{11}$, which is expected to contain both $Pr^{3+}$ and $Pr^{4+}$ the average shift obtained by fitting a single Lorentzian line is plotted.

and $PrO_2$ of about +0.6 mm/s is nearly as large as the shift between trivalent Pr and $PrO_2$. This indicates that for the $4f^1$ configuration of $Pr^{4+}$ covalency effects are more important than for $Pr^{3+}$. The isomer shift of $-0.82 \pm 0.06$ mm/s places Pr metal right into the middle of the trivalent region which suggests that the contribution of the conduction electrons to the s electron density at the nucleus in Pr is much smaller than in heavier trivalent rare earths like Sm and Dy [15,16]. A similar observation has recently been made for Nd [17]. Thus a strong decrease of the conduction electron densities seems to be a general feature in the light rare earths. The electron density D at the nuclei in tetravalent Pr ($4f^1$) is expected to be higher than in trivalent Pr ($4f^2$) due to the stronger shielding of the s electrons by the ($4f^2$) configuration. Scaling the electron densities in trivalent and divalent heavier rare earths [18] down with the relativistic factor [19], one obtains $D(4f^2) - D(4f^3) = 2.1 \times 10^{26}$ cm$^{-1}$ for Pr. Going one step further in the neglect of mutual shielding of the 4f electrons we assume $D(4f^1) - D(4f^2) \approx D(4f^2) - D(4f^3)$. Then from the isomer shift between $PrO_2$ and $Pr_2O_3$ we find $\delta<r^2> = +10 \times 10^{-3}$ fm$^2$ and from that between $Cs_2PrF_6$ and $K_3PrF_6$ the larger value $\delta<r^2> = +18 \times 10^{-3}$, the discrepancy reflecting the large difference between the isomer shifts of $PrO_2$ and the complex fluorides of tetravalent Pr. A comparison of the shifts of $Dy^{2+}$ and $Dy^{3+}$ [15] and of $Dy^{4+}$ in $CeO_2$ [20] with the present results suggests that the fluorides are the better representatives of $Pr^{3+}$ and $Pr^{4+}$ and that $\delta<r^2> = +18 \times 10^{-3}$ is the more trustworthy result. This $\delta<r^2>$ value has the same sign and a similar magnitude as $\delta<r^2>$ for the $1g_{7/2} - 2d_{5/2}$ single-hole transition in $^{151}$Eu [21].

We wish to thank Prof. R. Hoppe and Dr. R.H. Odenthal for making the $CsPrF_5$ and $Cs_2PrF_6$ sample available, and Dr. Ursel Zahn for her help in the preparation of several of the other scatterers.

REFERENCES

1. MORRISON, R.J., Thesis, University of Illinois, Urbana 1964, *Nucl. Sci. Abstr. 19*, 6638 (1965).

2.  Mac CHESNEY, J.B., WILLIAMS, H.J., SHERWOOD, R.C. and POTTER, J.F., *J. Chem. Phys. 41*, 3177 (1964).

3.  SHIRLEY, V.S. in *Hyperfine Structure and Nuclear Radiations*, E. Matthias and D. Shirley, eds. (North Holland Publishing Co., Amsterdam, 1968).

4.  COOK, D.D., PERSSON, B. and BENT, M., *Bull. Am. Phys. Soc. 14*, 1172 (1969).

5.  SIEGLAFF, C.L. and EYRING, L., *J. Am. Chem. Soc. 79*, 3024 (1957).

6.  CLIFFORD, A.F. and HUGHES, K.J. in *Proceedings of the 4th Conference on Rare Earth Research*, L. Eyring, ed. (Gordon and Breach, New York, 1964).

7.  SASTRY, R.L.N., MEHROTRA, P.N. and RAO, C.N.R., *J. Inorg. Nucl. Chem. 28*, 2167 (1966).

8.  ELLIOT, R.J., STEVENS, K.W.H., *Proc. Roy. Soc. A218*, 553 (1953).

9.  WATSON, R.E. and FREEMAN, A.J., in *Hyperfine Interactions*, A.J. Freeman and R.E. Watson, eds. (Academic Press, New York, 1967).

10. CABLE, J.W., MOON, R.M., KOEHLER, W.C. and WOLLAN, E.O., *Phys. Rev. Lett. 12*, 553 (1964).

11. HOLMSTRÖM, B., ANDERSON, A.C. and KRUSIUS, N., *Phys. Rev. 188*, 888 (1969).

12. GERTH, G., KIENLE, P. and LUCHNER, K., *Phys. Lett. 27A*, 557 (1968).

13. BERKOOZ, O., *J. Phys. Chem. Solids 30*, 1763 (1969).

14. HOPPE, R. and LIEBE, W., *Z. anorg. allg. Chem. 313*, 221 (1961).

15. HENNING, W., KAINDL, G., KIENLE, P., KÖRNER, H.J.,

KULZER, H., REHM, K.E. and EDELSTEIN, N., *Phys. Lett. 28A*, 209 (1968).

16. HENNING, W., BÄHRE, G. and KIENLE, P., contribution to this conference.

17. KAINDL, G., to be published.

18. HÜFNER, S. and PELZL, J., preprint, Freie Universität, Berlin, 1969.

19. SHIRLEY, D., *Rev. Mod. Phys. 36*, 339 (1964).

20. NOWIK, I., invited paper at this conference.

21. KIENLE, P., in *Hyperfine Structure and Nuclear Radiations*, E. Matthias and D. Shirley, eds. (North Holland Publishing Co., Amsterdam, 1968).

# EFFECTIVE MAGNETIC FIELDS AND ISOMER SHIFTS AT $^{61}$Ni NUCLEI IN Ni-Pd ALLOYS

F.E. OBENSHAIN, W. GLÄSER*, G. CZJZEK**, and J.E. TANSIL†

*Oak Ridge National Laboratory*††, *Oak Ridge, Tennessee, U.S.A.*

We have obtained nuclear gamma resonance absorption spectra of $^{61}$Ni in Ni-Pd alloys over the entire composition range with a source of $^{61}$Co embedded in a non-magnetic $^{64}$NiV(14%) foil. Source and absorber were held at 4.2°K. The magnetic hyperfine field at $^{61}$Ni in pure nickel is $H_{hf}$ = (-76 ± 1) kOe. Spectra taken with the alloy-absorbers show at every Pd concentration a distribution of hyperfine fields. The measurement <$|H_{hf}|$> has a minimum of (31 ± 1) kOe at 45 at. % Pd, a maximum of (173 ± 3) kOe at 90 at. % Pd. The behavior of the field for different concentrations can be explained if we assume that the hyperfine field at any $^{61}$Ni nucleus is determined by the distribution of nickel and palladium atoms on neighboring lattice sites, and that the palladium atoms give a strong positive contribution to the field. A correlation of our results with magnetic moments of Ni atoms in this alloy system has been established in the framework of the same model. Recoilless fractions for each concentration have been obtained.

---

*Exchange assignee from Kernforschungszentrum (KFK), Karlsruhe, Germany.
**Present address: Institut für Angewandte Kernphysik, Kernforschungszentrum Karlsruhe, Germany.
†Oak Ridge Graduate Fellow from the University of Tennessee.
††Operated by Union Carbide Corp. for the USAEC.

From this information it is deduced that approximately 75% of the observed energy shift for the Pd-rich alloys can be attributed to the second order Doppler shift. The residual shift $\sim(+8 \pm 5)$ µ/sec, which is the isomer shift, indicates that the $^{61}$Ni electronic configuration in Ni-Pd alloys changes only slightly over the entire concentration range.

REFERENCE

1. CABLE, J.W. and CHILD, H.R., *Bull. Am. Phys. Soc. 14*, 320 (1969).

# AVERAGE HYPERFINE FIELDS AT $^{106}$Pd NUCLEI IN Ni-Pd ALLOYS

M.M. EL-SHISHINI, R.W. LIDE, P.G. HURAY and J.O. THOMSON

*Department of Physics, University of Tennessee
Knoxville, Tennessee, USA*

The magnetic properties of Palladium metal and its ferromagnetic alloys have been extensively investigated in the last few years as examples of itinerant electron ferromagnets. To obtain information on electronic structure of Pd in such an alloy system, we have measured the average hyperfine magnetic fields for palladium in a series of ferromagnetic Ni-Pd alloys.

The measurements were carried out at 77°K by the IPAC technique using the 624-513 keV gamma ray cascade in $^{106}$Pd. The sources were prepared by diffusing $^{106}$Ru into the alloy foils for ∿50 hours at ∿1200°C in vacuum and then cooling rapidly. The sources were magnetized in a field of 10 kG using a magnet operating in liquid nitrogen. The source-magnet-dewar assembly was rotated slowly during the course of a measurement. Three 3" × 3" NaI(Tl) detectors located in the plane passing through the source and perpendicular to the applied field were employed, and two single channel analyzers were used with each detector, one set on each gamma ray. Four coincidence units with resolving times of the order of 60 nsec determined coincidences between detectors at angles of 150° and 160°.

To calculate the average fields in these alloys from the observed rotations of the correlation pattern we have used values for the intermediate state mean life $\tau$ = 18.4 psec [1] and for its gyromagnetic ratio g = 0.35 [2]. The observed fields have been corrected for

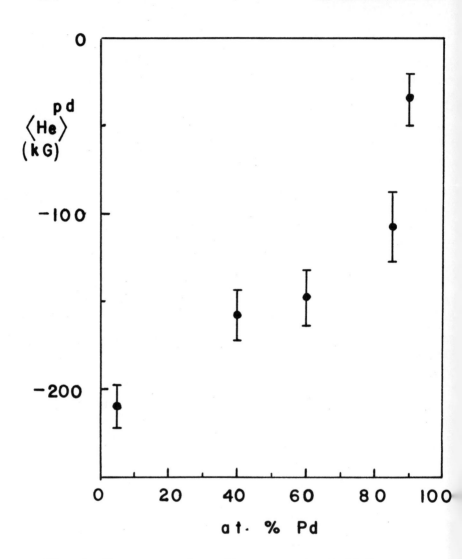

Fig. 1. The average hyperfine magnetic field for $^{106}$Pd$\langle H_{hf}^{Pd}\rangle$ as a function of composition for Ni-Pd alloys. The error bars represent the statistical error for each measurement.

# AVERAGE HF FIELDS AT $^{106}$Pd NUCLEI

the applied field and for the sample magnetization in order to obtain the hyperfine fields. In addition, the values of the hyperfine fields measured at 77°K have been extrapolated to 0°K assuming that $<H_{hf}{}^{Pd}>$ in these alloys scales with temperature in proportion to the saturation magnetization of nickel metal [3]. This correction amounts to no more than a few kG, and uncertainties in the extrapolation are thought to be negligible compared with our statistical error.

Fig. 1 shows our results for the average fields at 0°K for samples with Pd concentration ranging from 5 to 90% Pd. The result at 5% Pd is in agreement with the value of -185 ± 20 kG found using IPAC techniques by Murray et al. [2] for $^{106}$Pd in Ni metal, and the absolute value of 194 ± 4 kG from NMR studies [4] of dilute Pd in Ni. The values shown in Fig. 1 are average values of the magnetic field at a Pd nucleus, while the field at an individual Pd nucleus may depend strongly on the distribution of neighboring atoms. If the effects due to neighboring atoms in producing the hyperfine fields are additive, and if there is no short range order in these rapidly cooled alloys, then the average field measured here is an average over configurations of near neighbors containing the same percentage of Pd atoms as are present in the alloy. From Fig. 1, we see that these fields are negative over the whole concentration range studied. This is in contrast to the Ni hyperfine field [5,6], which changes sign near 45% Pd and rises to a large positive value at 90% Pd.

A part $H_{hf}{}^{(d)}$ of the average field at the Pd nuclei will arise from the d moment on the Pd atom. Cable and Child [7] have measured the individual atomic moments for the Ni and Pd atoms in a series of four Ni Pd alloys by neutron diffraction and scattering techniques. They find that the Pd d moment is 0.15 to 0.25 Bohr magnetons in the middle of the concentration range and decreases at either end. Using the g factor measurements of Fischer et al. [8] to find the net d-electron spin, S, and the value $H_{hf}{}^{(d)}$ = -689 S kG from Seitchik et al. [9], we estimate that $H_{hf}{}^{(d)} \simeq$ -45 to -70 kG between 50 and 75% Pd, and is smaller in magnitude for the remainder of the concentration range. We expect that the remaining large negative fields to be accounted for arise from interactions with neighboring atoms. From our data these contributions will be significant for Pd concentrations up to 85% Pd in Ni.

We wish to thank Dr. F.E. Obenshain and Dr. J.W. Cable for communicating some of their results and also for valuable discussions.

REFERENCES

1. STELSON, P.H. and GRODZINS, L., *Nucl. Data 1*, 21 (1965).

2. MURRAY, J., McMATH, T.A. and CAMERON, J.A., *Can. J. Phys. 45*, 1813 (1967).

3. WEISS, PIERRE and FORRER, R., *Annales de Physique* . *Ser. X, 5*, 153 (1926).

4. KONTANI, M. and ITOH, J., *J. Phys. Soc. Japan 22*, 345 (1967).

5. OBENSHAIN, F.E., GLÄSER, W., CZJZEK, G. and TANSIL, J.E., preceding paper at this conference and private communication.

6. ERICH, U., GÖRING, J., HÜFNER, S. and KANKELEIT, E., *Phys. Lett. 31A*, 492 (1970).

7. CABLE, J.W. and CHILD, H.R., *Bull. Am. Phys. Soc. 14*, 320 (1969), and private communication.

8. FISCHER, G., HERR, A. and MEYER, A.J.P., *J. Appl. Phys. 39*, 545 (1968).

9. SEITCHIK, J.A., GOSSARD, A.C. and JACCARINO, V., *Phys. Rev. 136*, A1119 (1964).

# STUDY OF HYPERFINE INTERACTIONS IN NEODYMIUM COMPOUNDS USING THE MÖSSBAUER EFFECT

F. T. PARKER and J. C. WALKER

*The Johns Hopkins University, Baltimore, Maryland*

Various neodymium compounds have been studied by the Mössbauer effect using the 72.5 keV transition in $^{145}$Nd [1]. The observed magnetic hyperfine fields are given in Table 1. Spectra for several of the compounds are shown in Fig. 1.

Cubic intermetallic rare earth compounds have received a great deal of attention over the past few years. The ionic properties of the rare earth constituent are determined by two terms in a crystal field expansion of the Coulombic potential, and by an exchange parameter. Esperiment usually yields but one number relating these three parameters, so that a unique determination of all three is difficult. Using a perturbation approach, Bleaney [2] estimated the magnetic moments to be expected in the $LnNi_2$ series. Recent Mössbauer data have shown saturation magnetization in the heavier rare earths in disagreement with the experimental and theoretical predictions. This has led some workers to conclude that a crystal distortion has brought a $|\pm J\rangle$ Kramers doublet lowest [3,4]. The neodymium studies reported here indicate that the heavy rare earth saturation fields can be explained on the basis of cubic symmetry.

For [100] quantization, the cubic point charge model crystalline potential can be written as

$$V = B_4[O_4^0 + 5O_4^4] + B_6[O_6^0 - O_6^4]$$

TABLE 1. - Observed Magnetic Hyperfine Fields in Neodymium Compounds

|  | $H_{eff}$, MOe | $H_r$ | $\mu_r$ |
|---|---|---|---|
| $NdAl_2$ | 2.57 ± 0.03 | 0.61 | 0.63[a] |
| $NdNi_2$ | 1.76 ± 0.05 | 0.42 | 0.55[a] |
| NdP | 2.66 ± 0.04 | 0.63 | 0.67[b] |
| NdAs | 3.19 ± 0.04 | 0.76 | 0.70[c] |
| NdSb | 2.92 ± 0.07 | 0.70 | 0.83[b] |
| NdBi | 2.95 ± 0.04 | 0.70 | 0.83[b] |
| $Nd_{1.5}Y_{1.5}IG$ | 2.05 ± 0.02 | 0.49 | 0.37[d] |
| $NdCl_3 \cdot 6H_2O$ | 1.57 ± 0.03 | 0.37 | 0.54[e] |

$H_r$ and $\mu_r$ are the reduced $H_{eff}$ and ionic moment, respectively.

(a) WALLACE, W. E., *Electronic Structure of Alloys and Intermetallic Compounds*, in *Progress in the Science and Technology of the Rare Earths*, Vol. 3 (Pergamon Press, Oxford, 1968), p. 1.

(b) TSUCHIDA, T. and WALLACE, W. E., *J. Chem. Phys. 43* 2885 (1965).

(c) TSUCHIDA, T., NAKAMURA, Y. and KANEKO, T., *J. Phys. Soc. of Japan, 26*, 284 (1969).

(d) GELLER, S., WILLIAMS, H. J. and SHERWOOD, R. C., *Phys. Rev. 123*, 1692 (1961).

(e) SCHULZ, M. B. and JEFFRIES, C. D., *Phys. Rev. 159*, 277 (1967).

TABLE 2. — Predictions of Reduced Rare Earth Hyperfine Fields in $DyAl_2$ and $ErAl_2$, Based on the Point Charge Model.

|  | Fourth order, $Z_1 = 0$ | | Fourth & sixth order, $Z_1 = 0$ | | Fourth & sixth order, $Z_1 = +1$ | | Fourth & sixth order, $Z_1 = -1$ | | Experiment |
|---|---|---|---|---|---|---|---|---|---|
|  | $H_{ex}$ const | $H_{ex}$ var'ble | $H_{ex}$ const | $H_{ex}$ var'ble | $H_{ex}$ const | $H_{ex}$ var'ble | $H_{ex}$ const | $H_{ex}$ var'ble |  |
| $Dy^{3+}$ | 0.99 | 1.00 | 1.00 | 1.00 | 1.00 | 1.00 | 1.00 | 1.00 | 1.00[a] |
| $Er^{3+}$ | 0.90 | 0.92 | 0.96 | 0.97 | 0.98 | 0.99 | 0.95 | 0.96 | 0.96±0.01[b] |
| $H_{ex}$,kOe | 110 | | 210 | | 170 | | 240 | | |

$H_{ex}$ is determined from the $NdAl_2$ measurement and the diagonalization, and corrected for a $J_{sf}^2 (g_J - 1) \langle J_z \rangle$ dependence.

(a) OFER, S., NOWIK, I. and COHEN, S. G. in *Chemical Applications of Mössbauer Spectroscopy* (Academic Press, New York, 1968), p. 434.

(b) PETRICH, G. and MÖSSBAUER, R. L., *Phys. Lett.* 26A, 403 (1968).

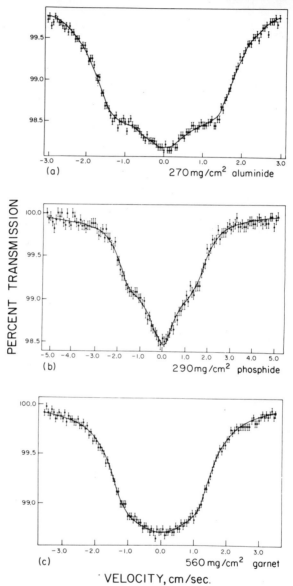

Fig. 1. Mössbauer Spectra in Some Neodymium Compounds
a. $NdF_3$ Source vs. $NdAl_2$ Absorber
b. $Sm_2O_3$ Source vs. $NdP$ Absorber
c. $Sm_2O_3$ Source vs. $Nd_{1.5}Y_{1.5}Fe_5O_{12}$ Absorber

where the bracketed quantities are angular momentum operators [5]. The quantities $B_4$ and $B_6$ are given in the point charge model by Bleaney. Assuming RKKY indirect exchange, the exchange interaction between two magnetic ions can be written [6]

$$H_{mag} = \frac{9\pi\xi^2}{4} \frac{J_{sf}^2(g_J - 1)^2}{V^2 E_F} F(2k_F R_{ij}) \vec{J}_i \cdot \vec{J}_j \quad (1)$$

with $\xi$ the number of free electrons per magnetic ion, V the atomic volume, $E_F$ the spherical Fermi energy, and $k_F$ the electronic wave vector at the Fermi surface. As Jones [7] has shown, for a spherical Fermi surface, F cannot vary with lattice constant.

The total Hamiltonian is usually written

$$H = 2(g_J - 1) \mu_\beta m_J H_{ex} + Z_2 V. \quad (2)$$

Comparison of Eq. (1) with Eq. (2) shows that $H_{ex}$ should vary approximately as $J_{sf}^2(g_J - 1)\langle J_z \rangle$. Eq. (2) was diagonalized for $Nd^{3+}$, $Dy^{3+}$ and $Er^{3+}$, the first two assuming [100] quantization and the latter [111] [8,5]. For $NdAl_2$ a point charge model calculation for pure fourth order crystal field ($Z_1 = 0$) yields $Z_2 B_4 \approx 4.5 \times 10^{-18}$ ergs and thus $H_{ex} \approx 110$ kOe. In Table II, predictions are given for reduced fields $H_r = H_{eff}/H_{free\ ion}$ for $DyAl_2$ and $ErAl_2$ on the basis of an $H_r$ of 0.60 for $NdAl_2$. Tabulated values [9,10,7] of $\langle r^4 \rangle$ and $\langle r^6 \rangle$, lattice constants, and $J_{sf}$ were used. The addition of a sixth order term increases the $Nd^{3+}$ $\Gamma_6$ (ground state) to $\Gamma_8$ (1) separation, requiring a larger exchange field to yield the observed hyperfine field; the $\Gamma_6 - \Gamma_8(3)$ separation in $Dy^{3+}$ and $\Gamma_8(2) - \Gamma_8(3)$ separation in $Er^{3+}$ then decrease. Various measurements on rare earth-aluminide compounds have provided the following values for $Z_1$: (a) susceptibility [11] with $Ce:LaAl_2$, $Z_1 \approx +1$ (for a calculated $Z_1 = 0$ overall splitting of 330°K); (b) neutron diffraction and specific heat [12] with $CeAl_2$, $Z \approx +2$; (c) NMR [13] with $TmAl_3$, $Z_1 \approx -1$. The predicted reduced fields agree well with experiment (conduction electron and core polarization have been neglected).

The rare-earth nickel compounds provide examples of both magnetization limits. In $NdNi_2$, $H_r$ is about 0.42, close to the minimum exchange field limit of 0.407. In DyNi$_2$, H$_r$ is approximately 1.0 [14], and in ErNi$_2$, 0.935 ± 0.045 [15]. The ErNi$_2$ result forces the field in DyNi$_2$ to be the free ion value, since even a small admixture of a sixth order term yields magnetic saturation values for Dy$^{3+}$. The prediction for NdNi$_2$ is H$_r$ = 0.48 or 0.45, assuming the mean value or lowest limit for the reduced field of ErNi$_2$ and Z$_1$ = 0.

The hyperfine fields with the III-V compounds agree fairly well with magnetization measurements. The Mössbauer spectra for all the intermetallics (except NdNi$_2$) exhibited an anomalous central peak indicative of some impurity phase [16]. With about 1/3 of the spectrum in the central peak, the most contaminated sample, NdSb, showed approximately the same field as seen in single crystal NdSb. Only in the NdNi$_2$ sample could the contamination be detected by X-ray powder diffraction (about 15%).

With the exchange field along the [111] axis, the garnet should show two magnetically inequivalent Nd$^{3+}$ sites of equal population. The spectrum could be fit well with just one hyperfine field. The chloride spectrum was fit under the assumptions of the effective field approximation ($g_1 \approx 0$) [17]. The garnet and chloride data indicate that the proportionality between ionic and nuclear fields does not hold for these compounds. J-mixing should not be expected for the chlorides because the crystal field splittings are quite small [17].

REFERENCES

1. KAINDL, G., *Phys. Lett. 28B*, 171 (1968).

2. BLEANEY, B., *Proc. Roy. Soc. (London) 276*, 28 (1963).

3. WIEDEMANN, W. and ZINN, W., *Phys. Lett. 24*, 506 (1967).

4. PETRICH, G. and MÖSSBAUER, R. L., *Phys. Lett. 26A*, 403 (1968).

. HUTCHINGS, M. T., *Point Charge Calculations of Energy*

*Levels of Magnetic Ions in Crystalline Electric Fields*, in *Solid State Physics*, 16 (Academic Press, New York, 1964).

6. DE GENNES, P., *C. R. Acad. Sc. (Paris) 247*, 1836 (1958).

7. JONES, E. D., *Phys. Rev. 180*, 455 (1969).

8. BOWDEN, G. J., BUNBURY, D., GUIMÃRAES, A. and SNYDER, R., *J. Phys. C. (Proc. Phys. Soc.) 1*, 1376 (1968).

9. FREEMAN, A. H. and WATSON, R. E., *Phys. Rev. 127*, 2058 (1962).

10. WERNICK, J. H. and GELLER, S., *Trans. A.I.M.E. 218*, 866 (1960).

11. WHITE, J. A., WILLIAMS, H. J. W., WERNICK, J. H. and SHERWOOD, R. C., *Phys. Rev. 131*, 1039 (1963).

12. HILL, R. W. and MACHADO DA SILVA, *Phys. Lett. 30A*, 13 (1969).

13. DE WIJN, H. H., VAN DIEPEN, A. M. and BUSCHOW, K.H.J., *Phys. Rev. 1*, 4203 (1970).

14. OFER, S., NOWIK, I. and COHEN, S. G. in *Chemical Applications of Mössbauer Spectroscopy* (Academic Press, New York, 1968), p. 469.

15. PETRICH, G., *Z. Physik 221*, 431 (1969).

16. LOH, E., PARKER, F. T. and WALKER, J. C., contribution to this conference.

17. SCHULZ, M. B. and JEFFRIES, C. D., *Phys. Rev. 159*, 277 (1967).

# ELECTRIC QUADRUPOLE INTERACTION OF $^{178}$Hf IN VARIOUS COMPLEX FLUORINE COMPOUNDS

E. GERDAU, B. SCHARNBERG and H. WINKLER

*II. Institut für Experimentalphysik, Universität Hamburg*

Mössbauer spectra of $^{178}$Hf in polycrystalline $A_2HfF_6$ and $A_3HfF_7$ compounds (A for alkali cation) were measured at 4.2°K using the 93 keV radiation of a $^{178}$W source in W metal.

A. *Electric Quadrupole Interaction.*

The measured spectra can be interpreted by an axial asymmetric electric quadrupole interaction. Both the interaction strength $eQV_{zz}/4I(2I - 1)$ and the axial asymmetry $\eta = (V_{xx} - V_{yy})/V_{zz}$ are influenced by the special cation (Table 1). With the $NH_4$-cation two crystalline modifications of the hexafluoride can be distinguished by their different interactions (Fig. 1). The modification I and the other hexafluorides show a strong anisotropy with an $\eta$ near 1. This means that there exist two components of the EFG tensor - $V_{zz}$ and $V_{yy}$ in the usual nomenclature - which have nearly equal absolute values but different signs, while the third component - $V_{xx}$ - is close to zero.

In the case of $K_2HfF_6$ (Fig. 2) the atomic positions of all constituents are known [1]. The crystalline structure is pseudohexagonal with hafnium fluorine chains in the direction of the crystalline c axis. Numerical calculations with point charges fail by more than one order of magnitude in the absolute value of the interaction. This could be explained by Sternheimer antishielding corrections. It seems probable from these calculations that the strong positive component of the EFG tensor is caused by

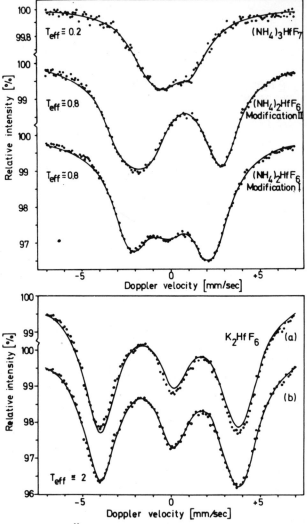

Fig. 1. Above: Mössbauer spectra of the two modifications of $(NH_4)_2HfF_6$ and of $(NH_4)_3HfF_7$.

Below: Spectrum of $K_2HfF_6$. (a) Least-squares-fit without Goldanskii-Karyagin effect, the intensities were not reproduced in the right proportions. (b) Goldanskii-Karyagin effect was taken into account.

ELECTRIC QUADRUPOLE INTERACTION OF $^{178}$Hf

TABLE 1. – Interaction Constants and Anisotropy Parameters of the Debye-Waller Factor

| | $\frac{1}{24} eQV_{zz}$ [$10^{-6}$ eV] | $\eta$ | $f_0$ [%] | $b_2$ | $b_4$ |
|---|---|---|---|---|---|
| Li$_2$HfF$_6$ | +0.139(4) | 0.79(4) | 8.1(8) | +0.2(2) | 0.4(3) |
| Na$_2$HfF$_6$ | +0.143(4) | 0.80(2) | 11.1(8) | +0.17(2) | 0 |
| K$_2$HfF$_6$ | −0.192(5) | 0.89(2) | 16.5(9) | +0.40(4) | 0.10(5) |
| Rb$_2$HfF$_6$ $^{a,b}$ | 0.145(15) | 0.9...1 | − | − | − |
| Cs$_2$HfF$_6$ $^b$ | 0.148(4) | 0.99(4) | 14.7(1.5) | +0.27(3) | 0 |
| (NH$_4$)$_2$HfF$_6$ (I) | +0.109(4) | 0.92(2) | 11.8(8) | −0.95(9) | 0.5(2) |
| (II) | +0.149(5) | 0.33(3) | 11.5(2.5) | −0.2(1) | 0.2(1) |
| Li$_3$HfF$_7$ | +0.039(1) | 0 | − | − | − |
| Na$_3$HfF$_7$ | +0.056(2) | 0 | − | − | − |
| K$_3$HfF$_7$ | −0.048(2) | 0 | − | − | − |
| Rb$_3$HfF$_7$ | −0.083(2) | 0 | − | − | − |
| Cs$_3$HfF$_7$ | −0.077(2) | 0 | − | − | − |
| (NH$_4$)$_3$HfF$_7$ | +0.058(2) | 0 | − | − | − |

a. Slight admixture of heptafluoride.
b. The sign of the interaction is undefined because $\eta \cong 1$.

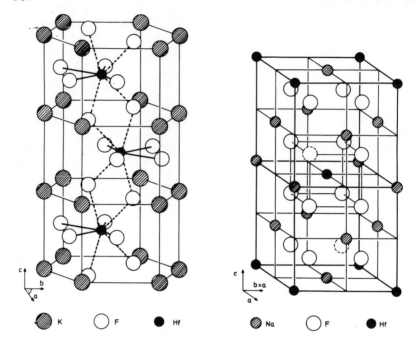

Fig. 2. Left: Crystalline structure of $K_2HfF_6$.

Right: Crystalline structure of $Na_3HfF_7$. Every eighth flourine position is empty.

the $Hf^{4+}$ neighbours in the c direction, whereas one strong negative and one nearly vanishing component in the (a,b)-plane are not reproduced. Obviously the point charge model is too simple and covalent bonding effects between Hf and the surrounding F should be taken into account.

Similar arguments may be right for the other hexafluoride compounds. Regarding the negative sign of the nuclear quadrupole moment [2] ($Q(^{178}Hf)$ = -1.93(5) b), Table 1 shows that in the case of small cations the negative component of the EFG tensor outweighs the positive one.

From the heptafluorides the Na compound is known to be isostructural with $\alpha-Na_3UF_7$ (Fig. 2) [3]. The lattice is body-centered tetragonal (with $c \cong 2a$), but every

eighth fluorine position is statistically empty. The
interaction might be caused by this hole. The exact coordination of the fluorine ions in other compounds is not
known.

B. *Goldanskii-Karyagin Effect*

The theoretical curves in Fig. 1 were calculated by
numerical evaluation of the transmission integral. In
order to get satisfying fits of the spectra one has to
take into account the spatial anisotropy of the Debye-
Waller factor. This was done by [4]

$$f(\theta,\phi) = \frac{1}{4\pi}\exp\{-\frac{1}{\lambda^2}[(<x_x^2>\cos^2\phi + <x_y^2>\sin^2\phi)\sin^2\theta + <x_z^2>\cos^2\theta]\}$$

assuming that the off-diagonal elements $<x_i x_k>$ vanish in
the coordinate system in which the interaction Hamiltonian
is diagonal.

In the case of $I_e = 2^+$ and $I_g = 0^+$ the intensity of
the different transitions $I_{n_e n_g}$ between the sublevels is
given by [5]

$$I_{n_e 0}(\theta,\phi) = f(\theta,\phi)\sum_{m_e m_e'} <E_{n_e}|m_e><E_{n_e}|m_e'>^* F_{m_e m_e'}^{22}(\theta,\phi)$$

with

$$F_{MM'}^{LL'}(\theta,\phi) = \sum_{k=even} (-1)^{M'+1}[(2L+1)(2L'+1)]^{-\frac{1}{2}} \times$$

$$C(LL'k, M-M'M-M')C(LL'k, 1-10)D_{M-M'0}^{k*}(z \to \vec{k})$$

$<E_n|m>$ are the elements of the unitary transformation
matrix which diagonalizes the electric quadrupole interaction Hamiltonian. For polycrystalline absorbers one has
to integrate over all angles $\theta$ and $\phi$. Integration over $\phi$
yields [6]

$$I_{n_e 0}(\theta) = \sum_{m_e} |<m_e|E_{n_e}>|^2 f(\theta) F_{m_e m_e}^{22}(\theta)$$

where

and
$$f(\theta) = \tfrac{1}{2}f \exp(\delta \sin^2\theta) I_0\left(\tfrac{i}{2}\delta \sin^2\theta\right)$$

$$f = \exp\{-\tfrac{1}{\lambda^2}\langle x_z^2\rangle\}, \quad \delta = \tfrac{1}{\lambda^2}|\langle x_x^2\rangle - \langle x_y^2\rangle|,$$

$$\delta = \tfrac{1}{\lambda^2}\left(\langle x_z^2\rangle - \tfrac{\langle x_x^2\rangle + \langle x_y^2\rangle}{2}\right)$$

With the expansion $f(\theta) = -f\left(1 + \sum_{n=1}^{\infty} b_n P_n(\cos\theta)\right)$ the integration can be performed because the $F_{MM}^{22}$ are linear combinations of $P_0$, $P_2$ and $P_4$. Using the orthogonality relations of the Legendre polynominals one gets

$$f_0 = -\tfrac{1}{2}f \int \exp(\delta \sin^2\theta) I_0\left(\tfrac{i}{2}\delta \sin^2\theta\right) d\cos\theta$$

$$b_n = \tfrac{2n+1}{2} \cdot \tfrac{f}{f_0} \int \exp(\delta \sin^2\theta) I_0\left(\tfrac{i}{2}\delta \sin^2\theta\right) P_n(\cos\theta) d\cos\theta$$

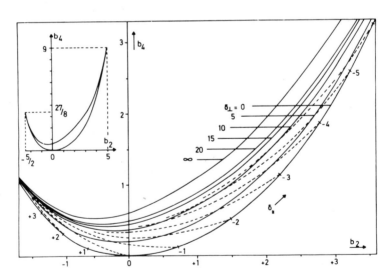

Fig. 3. Range of the allowed $(b_2, b_4)$ pairs. The attached $(\delta_\parallel, \delta_\perp)$ values are indicated.

ELECTRIC QUADRUPOLE INTERACTION OF $^{178}$Hf

The $b_n$ are determined exclusively by $\delta_\parallel$ and $\delta_\perp$, i.e. by the deviations from isotropy. The fit yields values for $f_0$, $b_2$ and $b_4$. Fig. 3 shows how $\delta_\parallel$ and $\delta_\perp$ are attached to the parameters $b_2$ and $b_4$ and gives the range of the allowed ($b_2, b_4$) pairs. The results for $b_2$ and $b_4$ given in Table 1 seem to be reasonable as they lie in the allowed range.

REFERENCES

1. *Struct. Rep. 20*, 224 (1956); BODE, H. and TEUFER, G., *Acta Cryst. 9*, 929 (1956), and *Z.f.anorg.Ch. 283*, 18 (1956).

2. BOOLCHAND, P., ROBINSON, B. L. and JHA, S., *Phys. Rev. 187*, 475 (1969).

3. *Struct. Rep. 23*, 291 (1959), and *11*, 333 (1947/48).

4. GOLDANSKII, V. I. and MAKAROV, E. F. in *Chemical Applications of Mössbauer Spectroscopy*, ed. by V. I. Goldanskii and R. H. Herber, Academic Press, New York and London, 1968, p. 104.

5. FRAUENFELDER, H. and STEFFEN, R. M. in *Alpha-, Beta- and Gamma-Ray Spectroscopy* vol. 2, ed. by K. Siegbahn, North-Holland Publ. Comp., Amsterdam, 1965.

6. ABRAMOWITZ, M. and STEGUN, I. A., *Handbook of Mathematical Functions*, Dover Publ. Inc., New York, 1965, p. 376.

# ANGULAR DISTRIBUTION, LINE POSITION AND LINE WIDTH OF 14.4 Kev RADIATION IN Fe-Pd ALLOYS

G.R. ISAAK and U. ISAAK

*Department of Physics, University of Birmingham, Birmingham, England*


Hyperfine interactions at Fe sites in Fe-Pd alloys containing 1.7, 6.1, 8.9 and 11.8 at.% Fe were investigated at 300°K by measuring the angular distribution of the resonantly scattered radiation from a $^{57}$Co in Pd source. The measured asymmetries W(135°)/W(90°) for these alloys were 1.197 ± 0.008, 1.160 ± 0.007, 1.123 ± 0.007 and 1.080 ± 0.006 compared with the unattenuated value of 1.214.

These results are interpreted in terms of electron paramagnetic relaxation times ranging from 1 $\times$ 10$^{-12}$/S(S + 1) to 14 $\times$ 10$^{-12}$/S(S + 1) seconds for the various alloys where S is the spin on the Fe site. On the assumption of a constant spin S the relaxation times vary quadratically with concentration (Fig. 1). The relaxation times also vary quadratically with the Curie temperature of the alloy [1].

The corresponding full widths at half height of the transmission spectra extrapolated to zero thickness of the absorber were 0.321 ± 0.003, 0.347 ± 0.001, 0.375 ± 0.002 and 0.431 ± 0.002 mm s$^{-1}$.

If one subtracts from the transmission widths the source width on the assumption of Lorentzian shapes (Fig. 2a) and allows for the finite broadening due to the above mentioned relaxation effects (assuming S = $\frac{1}{2}$,

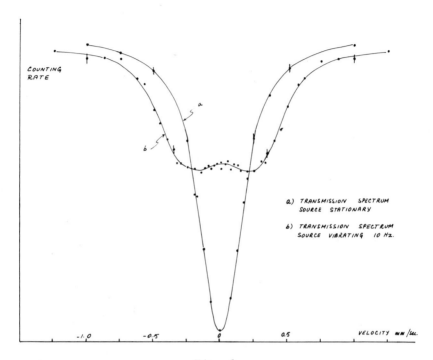

Fig. 1.

say) a large residual width remains which is ascribed to a distribution of isomeric shifts. This distribution appears to increase linearly with concentration (Fig. 2b) and is at least in this respect consistent with the expectation of the iron atoms being located on the spatially fluctuating electron density oscillations surrounding each impurity atom (Friedel effect).

The corresponding absorption line energies decreased with concentration and were +3.7 ± 0.7, -5.0 ± 0.7, -6.2 ± 0.8 and -2.2 ± 0.9, all in fractional parts in $10^{15}$, measured relative to the source. These measurements were carried out near the steepest part of the transmission lines. These shifts appear to depart from the line joining pure Pd with pure Fe by a positive energy shift which increases quadratically with increasing concentration. This shift may well have a

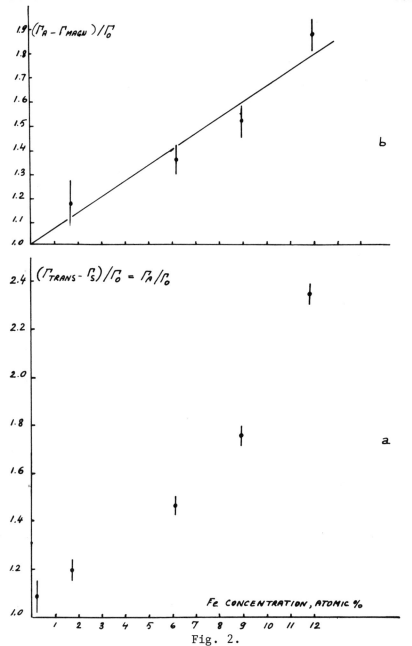

Fig. 2.

magnetic origin and may be similar to the shift observed by Kerler et al. [2].

REFERENCES

1.  TROUSDALE, W.L. et al., *J. Appl. Phys.* *38*, 922, (1967).

2.  KERLER, W. et al., *Z. Physik* *173*, 321 (1963).

# STUDIES OF MÖSSBAUER FRACTIONS AND HFS IN FROZEN SOLUTIONS[†]

A. SIMOPOULOS, H.H. WICKMAN[*], D. PETRIDIS and A. KOSTIKAS

*Nuclear Research Center "Democritos",
Athens, Greece*

An interesting question raised by recent studies in frozen aqueous solutions (FAS) is the nature of the "solid state" transition leading to pronounced decrease or loss of the recoil free fraction (rff) at a temperature characteristic of the particular solvent/solute system [1,2,3]. A crystallographic transformation [1,2] and a local process [3] have been suggested as possible mechanisms for the rff change. A third possibility is that the metal ions are initially frozen in a vetreous region and that the observed loss of rff is associated with a glass transition. With this in mind we have carried out preliminary experiments in two systems which show clear glassy transitions followed by crystallization transfromations, the object being to observe the coupling between Mössbauer parameters and the characteristic temperature associated with the glass transformation.

The solute/solvent systems were suggested by EPR studies of Allen [4] which show definite "before" and "after" effects in the crystal field splitting of $Mn^{2+}$ in $MnCl_2$/12N HCl and $MnCl_2$/methanol frozen solutions.

The systems chosen for the Mössbauer experiments

---

[†] Supported in part by NATO Research Grant No. 407
[*] On leave from Bell Telephone Laboratories, U.S.A.

were $FeCl_2$ and $FeCl_3$ enriched in $^{57}Fe$ in AR grade methanol or 12N HCl. Iron concentrations were of the order of or less than one atomic percent. The initial state of the samples was always a clear cracked glass at 77°K. This changes irreversibly to an opaque polycrystalline phase at typical temperatures of 119°K (methanol) and 146°K (12N HCl). The crystallization temperature $T_c$ varies with the heating rate, shifting towards higher values the larger the rate.

It is a reflection of the uncertain and complicated chemistry of these "simple" species in solution that a rather broad and incompletely understood range of Mössbauer spectra have been found. The data are summarized here according to the ion/solvent system. All observations initially refer to the passage from glassy to opaque phase at the characteristic temperature $T_c$ noted above.

*$Fe^{2+}$, $Fe^{3+}$/Methanol:* A pronounced reduction or complete loss of the rff was observed for both ferrous and ferric solutes. A possible reason for this behaviour, based on relaxation processes at the glass transition, is given elsewhere [5]. The quadrupole splitting of the ferrous species decreases from 3.42 mm/sec to 3.29 mm/sec.

The ferric species display paramagnetic hfs ($H_{eff}$ = 530 kOe) which is different in the two phases (Fig. 1) suggesting a longer spin-lattice relaxation time and a larger crystal field splitting in the glassy phase.

*$Fe^{2+}$, $Fe^{3+}$/12 N HCl:* The ferrous species display a quadrupole doublet of 3.6 mm/sec, which disappears at $T_c$ = 146.5°K. It reappears within one hour together with a new quadrupole doublet of 1.5 mm/sec. The latter grows up gradually at the expense of the former.

The ferric species display paramagnetic hfs ($H_{eff}$ = 530 kOe) broadened by spin-lattice relaxation near $T_c$. At an abrupt change to a single broad (1mm/sec) line occur. Fig. 2 shows Mössbauer spectra at 4.2°k for the glassy and polycrystalline states. Application of a 5

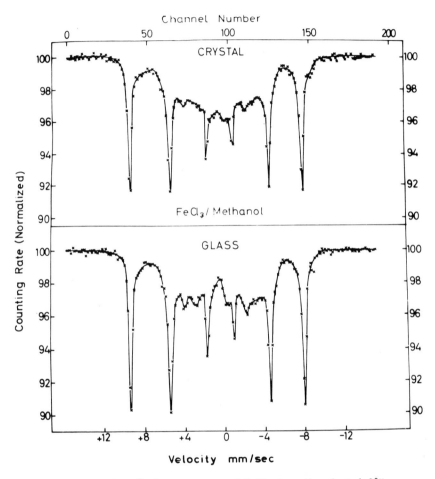

Fig. 1. Mössbauer spectra of FeCl₃ in methanol at 4.2°K.

kOe field at 4.2°K broadens the line but does not produce a well defined magnetic hfs. The overall width corresponds to a field of approximately 95 kOe, a value in reasonable agreement with that expected from a Brillouin function $B_{5/2}$, with saturation field of 530 kOe.

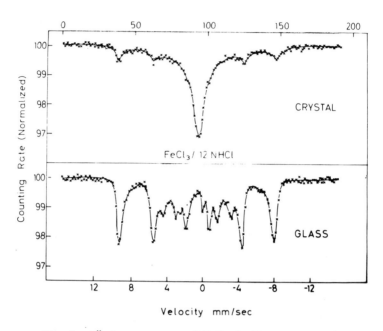

Fig. 2. Mössbauer spectra of $FeCl_3$ in 12 N HCl at 4.2°K.

*Summary*. The data indicate a correlation between the bulk glassy to polycrystalline transformation and an enhanced mobility of the Fe ions in frozen methanol and 12 N HCl solutions. In $Fe^{3+}$/12 N HCl the drastic change of the paramagnetic hfs at $T_c$ indicates some aggregation of the $Fe^{3+}$ ions. The $Fe^{2+}$/12 N HCl data are difficult to interpret, but clearly imply a phase separation upon crystallization.

REFERENCES

1. DEZSI, I., KESZTHELYI, L., POCS, L. and KORECZ, *Phys. Lett.* 14, 14 (1965); DEZSI, I., KESZTHELYI, L., MOLNAR, B. and POCS, L., *Hyperfine Structure and Nuclear Radiations*, Matthias and Shirley, Eds., p. 566 (1967).

2. NOZIK, A. and KAPLAN, M., *J. Chem. Phys. 47*, 2960 (1967), DILORENZO, J.V. and KAPLAN, M., *Chem. Phys. Lett. 3*, 216 (1969).

3. PELAH, I. and RUBY, S.L., *J. Chem. Phys. 51*, 383 (1969).

4. ALLEN, B.T., *J. Chem. Phys. 43*, 3820 (1965).

5. WICKMAN, H.H., Relaxation effects in Mössbauer Spectra. These proceedings.

# HYPERFINE INTERACTIONS IN $Yb_6Fe_{23}$

G. GORETZKI, G. CRECELIUS and S. HÜFNER

*IV. Physikalisches Institut, Freie Universität Berlin*

Ytterbium and Europium are the only rare earth elements being divalent in most metallic environments. Divalent Ytterbium (ground state $^1S_0$) is nonmagnetic as, e.g., in Yb metal. Sizable hyperfine fields at the Yb nucleus in a metallic environment can be produced by implanting Yb into a ferromagnetic host. Because of the marked differences in the ionic radii between Fe and Yb, it is questionable whether the implanted ions really come to rest at regular lattice sites. The only experiment reported so far by Boehm et al. [1] using this technique yields $H_{Yb}(300°K) = (720 ± 240)$ kOe. Therefore we investigated the iron-rich intermetallic compound $Yb_6Fe_{23}$. The compound can only be produced using powder-metallurgical methods. An X-ray analysis of our sample showed the expected pattern ($Yb_6Mn_{23}$ structure) and only small contaminations of α-iron. The hyperfine fields were measured using a conventional transmission Mössbauer setup. The high temperature measurements were performed in a resistance furnace with the temperature stabilized to 0.25°K. The temperature calibration is accurate to ±5°K. The results for the iron measurements taken with a source of $^{57}Co$ in Pd are

$H_{Fe}(300°K) = (325 ± 5)$ kOe
$T_c = (1018 ± 5)$ °K

as compared to

$H_{Fe} = 330$ kOe
$T_c = 1044$ °K

for iron metal.

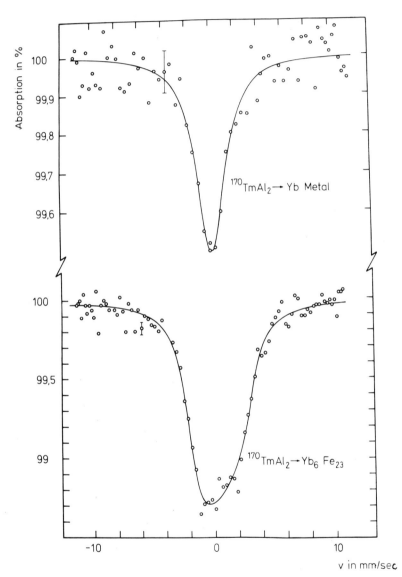

Fig. 1. Resonance absorption of the 84-keV γ-line of $^{170}$Yb.

The Yb measurements were performed using the 84-keV γ-line of $^{170}$Yb in TmAl$_2$. A spectrum taken at 4.2°K with an absorber of Yb$_6$Fe$_{23}$ is presented in the figure. The solid line is the result of a least squares fit yielding

$$H_{Yb}(4.2°K) = (258 \pm 7) \text{ kOe}$$
$$\Delta = (-0.31 \pm 0.03) \text{ mm/sec}$$
$$eQV_{zz}/4 = (0.13 \pm 0.03) \text{ mm/sec.}$$

The isomer shift is more negative than that in Yb metal (-0.08 mm/sec) but definitely different from that in any trivalent compound which is positive, thus indicating that Yb is divalent in Yb$_6$Fe$_{23}$.

# REFERENCES

1. BOEHM, F., HAGEMANN, G.K. and WINTHER, A., *Phys. Lett.* 21, 217 (1966).

# MÖSSBAUER SPECTROSCOPIC INVESTIGATION OF THE REVERSIBLE OXYGENATION OF HEMOGLOBIN

A. TRAUTWEIN

*Institut für Metallphysik, Universität des Saarlandes, Saarbrücken, Germany*

P. SCHRETZMANN

*Chemisches Institut, Universität Tübingen, Germany*

The binding of molecular oxygen to the iron in hemoglobin has been the subject of intensive investigations since the early efforts of Pauling and Coryell in 1936 [1]. But the nature of the oxygen coordination is not yet fully understood. The Mössbauer technique has become a useful tool in the study of metal-ligand interactions in biological iron complexes. An analysis of our Mössbauer spectroscopic results should throw some light on the problem of reversible oxygenation of hemoglobin.

*Results*

1. It was shown, that there exists an intermediate spin state $^3E(S = 1)$, only 250 $cm^{-1}$ above the high spin ground state $^5B_2(S = 2)$ of the iron (II) in deoxyhemoglobin (Hb), both states interacting via spin-orbit coupling [2-5]. The many electron term $^3E$ in Hb therefore is populated at physiological temperatures to a considerable extent.

2. The iron in oxyhemoglobin (HbO$_2$) is diamagnetic. Oxygen binding to Hb causes a negative electric field gradient (EFG) and a relatively small asymmetry parameter $\eta \sim 0.3$ [2-5].

These results are in agreement with the assumption that the interaction between the spin triplet of oxygen (S = 1) and the intermediate spin state $^3E$ of iron in Hb would generate a molecular orbital (m.o.) of zero net spin in HbO$_2$ lying below the $^5B_2$ ground state of Hb [2-5].

*Structural and Kinetic Implications*

O$_2$ is known to form antiferromagnetic coupled molecule complexes in its triplet ground state $^3\Sigma_g^-$, e.g. (O$_2$)$_2$ [6] or (O$_2$) in solid oxygen [7]. One can expect that antiferromagnetic spin coupling between the O$_2$ molecule in its $^3\Sigma_g^-$ ground state and the iron of Hb varies the thermal equilibrium between the $^3E$ triplet state. Antiferromagnetic coupling between O$_2$ and the heme iron, which does not imply valence bonds, may produce a small increase in the heme plane ligand field, as the iron, pentacoordinated in Hb, approximates hexacoordination during the approach of O$_2$. It was shown (see formulas 32 a-c in our earlier paper [2]) that an increased heme plane ligand field lowers the energetic gap between $^3E$ and $^5B_2$. Therefore one can suppose, that antiferromagnetic spin coupling results in considerable spin quenching and some energetic stabilization:

$$Fe(II)(^5B_2 \rightleftharpoons {}^3E) + (^3\Sigma_g^-)O_2 \rightarrow Fe(II)(^5B_2 \rightleftharpoons {}^3E) \cdots (^3\Sigma_g^-)O_2$$
$$S \approx 1 (\downarrow\downarrow) \qquad S = 1 (\uparrow\uparrow)$$

It may initiate the oxygenation, however it seems not to be sufficient to explain the thermodynamically stable state of HbO$_2$. The latter is usually described (for a review see [8]) by a double-bond system, exhibiting either a bent end-on structure as proposed by Pauling [9] or a structure with oxygen parallel to heme as proposed by Griffith [10].

To explain a valence-bound stucture with zero net spin, Griffith assumed an electronic rearrangement in the $O_2$ molecule ($S = 1 \rightarrow S = 0$) as well as in the iron ($S = 2 \rightarrow S = 0$) prior to oxygenation, but "it is not suggested that the actual reaction of Hb with $O_2$ may follow this course" [10].

An involvement of iron in the intermediate spin state $^3E(S = 1)(d_{xy}(2), d_{xz}(2), d_{yz}(1), d_{z^2}(1), d_{x^2-y^2}(0)$ restricts the electronic rearrangement preceding the formation of a double bond to the oxygen molecule, which possesses metastable excited triplet states [11].

There are two possibilities for such an electronic rearrangement and subsequent bonding to the $^3E$ iron (II):

a) A $\pi - \pi^x$ transition within the $O_2$ molecule creates a singly populated $\pi$-bonding m.o., which can combine with the singly populated $d_{z^2}$ orbital of the iron in the $^3E$ state forming a $\sigma$-bonding m.o. between the iron and the $O_2$ group according to the geometry proposed by Griffith (Fig. 1).

b) A $n - \pi^x$ transition within the $O_2$ molecule results in a singly populated non-bonding m.o. on the $O_2$ group, which can combine with the $d_{z^2}$ orbital of the heme iron in the $^3E$ state, causing a $\sigma$-bond with the geometry proposed by Pauling (Fig. 2).

In both cases, additional $\pi$-bonds can be formed between an electron in the $d_{yz}$ orbital of the iron and an electron in a $\pi^x$-orbital of the $O_2$ group. Both classical models for the oxygen-hemoglobin bonding can be formed by combining iron (II) in the $^3E$ spin state with excited triplet states of the $O_2$ molecule. They can be formed without alterations of the spin multiplicity of the components during the formation or the break up of the oxygenated adduct:

$$Fe(II)(^5B_2 \rightleftharpoons ^3E) \cdots (^3\Sigma_g^-)O_2 \rightleftharpoons Fe \begin{pmatrix} \downarrow\uparrow \\ \downarrow\uparrow \end{pmatrix} O_2$$
$$S \approx 1 (\downarrow\downarrow) \qquad S = 1 (\uparrow\uparrow) \qquad S = 0$$

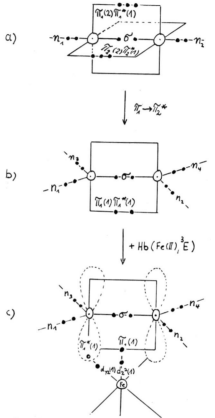

Fig. 1. Hypothetical $O_2$ bonding in hemoglobin (Griffith geometry):
a) free $O_2$ molecule, $^3\Sigma_g^-$ ground state, $S = 1$
b) $O_2$ molecule after promotion of an electron from $\pi_1$ to $\pi_2^*$, $S = 1$
c) fixation of the $\pi_1 \to \pi_2^*$ excited $O_2$ to heme iron in the intermediate spin state $^3E$ ($d_{xy}(2)$, $d_{xz}(2)$, $d_{yz}(1)$, $d_{z^2}(1)$, $d_{x^2-y^2}(0)$), $S = 1$

$\sigma$-bonding m.o.: $d_{z^2}(1) + \pi_1(1)$
$\pi$-bonding m.o.: $d_{yz}(1) + \pi_1^*(1)$
(s-electrons of $O_2$ neglected).

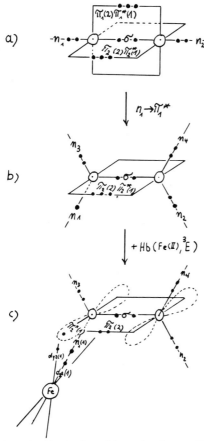

Fig. 2. Hypothetical $O_2$ bonding in hemoglobin
(Pauling geometry):
a) free $O_2$ molecule, $^3\Sigma_g^-$ ground state, $S = 1$
b) $O_2$ molecule after promotion of an electron from $n_1$ to $\pi_1^*$, $S = 1$
c) fixation of the $n_1 \rightarrow \pi_1^*$ excited $O_2$ to heme iron in the intermediate, spin state $^3E$, $S = 1$

$\sigma$-bonding m.o.: $d_{z^2}(1) + n_1(1)$
$\pi$-bonding m.o.: $d_{yz}(1) + \pi_2^*(1)$
(s-electrons of $O_2$ neglected).

This conservation of the spin multiplicity may be the crucial point for the easy kinetic reversibility of the oxygenation. The theory mentioned above is in accord with the proposal of Mason [12] that the bond length in transition metal-$O_2$-systems are best explained assuming higher triplet states of $O_2$.

Lang explains the magnitude and the sign of the EFG in $HbO_2$ under the a priori assumption of the Griffith geometry [13,14]. However we think that neither the magnitude nor the sign of the EFG tensor exclude the Pauling geometry, which is in accord with the relatively small asymmetry parameter $\eta$, the preliminary X-ray diffraction results at oxymyoglobin [15], and the preliminary IR data at $HbO_2$ [18].

The electron density distribution within the $FeO_2$ group of $HbO_2$ is not yet known. The chemical shifts found in $HbO_2$ indicate however, that the electron density at the iron nucleus is comparable to the electron density in some tetragonal distorted iron (II) complexes with intermediate spin $S = 1$ [4,16,17]. A formal valence state Fe (II) in $HbO_2$ should therefore be preferred to Fe (III).

Further investigations concerning these problems are in progress.

REFERENCES

1. PAULING, L. and CORYELL, C.D., *Proc. Nat. Acad. Sci. USA 22*, 210 (1936).

2. EICHER, H. and TRAUTWEIN, A., *J. Chem. Phys. 50*, 2540 (1969).

3. EICHER, H. and TRAUTWEIN, A., *J. Chem. Phys. 52*, 932 (1970).

4. TRAUTWEIN, A., EICHER, H. and MAYER, A., *J. Chem. Phys. 52*, 2473 (1970).

5. TRAUTWEIN, A., EICHER, H., MAYER, A., ALFSEN, A., WAKS, M., ROSA, J. and BEUZARD, Y., *J. Chem. Phys.*, to appear.

6. ARNOLD, S., FINLAYSON, N. and OGRYZLO, E.A., *J. Chem. Phys. 44*, 2529 (1966).

7. BARRETT, C., *J. Chem. Phys. 47*, 592 (1967).

8. BAYER, E. and SCHRETZMANN, P., p. 181 in *Structure and Bonding*, Vol. 2, C.K. Jørgensen, ed., Springer-Verlag Berlin, 1967.

9. PAULING, L., *Nature 203*, 182 (1964).

10. GRIFFITH, J.S., *Proc. Roy. Soc. (London) A 235*, 23 (1956).

11. HERZBERG, G., *Spectra of Diatomic Molecules*, p. 446, Van Nostrand Co., Princeton, 1957.

12. MASON, R., *Nature 217*, 543 (1968).

13. LANG, F. and MARSHALL, W., *Proc. Phys. Soc. (London) 87*, 3 (1966).

14. LANG, G., *Proceedings of the Conference on Application of the Mössbauer Effect*, Tihany, Hungary, June 1969.

15. WATSON, H.C. and NOBBS, C.L., p. 37 in *Biochemie des Sauerstoffs*, B. Hess and Hj. Staudinger, eds, Springer-Verlag, Berlin, 1968.

16. DALE, B.W., WILLIAMS, R.J.P., EDWARDS, P.R. and JOHNSON, C.E., *J. Chem. Phys. 49*, 3445 (1968).

17. KÖNIG, E. and MADEJA, K., *Inorganic Chemistry 7*, 1849 (1968).

18. McCOY, S. and CAUGHEY, W., *Biochemistry 9*, 2392 (1970).

# TEMPERATURE DEPENDENCE OF HYPERFINE FIELDS IN $DyCo_2$, $DyNi_2$ AND Dy-METAL

E. LOH

*Towson State College,
Baltimore, Maryland*

F.T. PARKER and J.C. WALKER

*The Johns Hopkins University,
Baltimore, Maryland*

In many recent Mössbauer experiments on the temperature dependence of the hyperfine field at rare-earth nuclei in magnetically ordered compounds, strong central peaks and a general smearing of the spectra occur as the temperature of the sample is raised toward the critical point. Most explanations of these phenomena involve relaxation processes [1]. In iron compounds, similar effects have been explained on the basis of critical superparamagnetism [2], with small volumes of the sample undergoing spin-flips. Superparamagnetic phenomena should be observable for only a small temperature span. Since macroscopic measurements involve a time interval long compared to the Mössbauer experiment, the effective magnetization of the sample should be near zero when superparamagnetic relaxation becomes visible in the Mössbauer spectrum. This is not observed to be the case in the rare earths.

In this paper, we report temperature studies of the Mössbauer effect in $DyCo_2$, $DyNi_2$, and Dy-metal. The data do not agree with other work [3] on these materials, and indicate a sample dependence of the "relaxation" process. A possible explanation for the apparent paramagnetic contamination can be given in terms of regions of differing critical temperature. Qualitative agreement with the present $DyCo_2$ data could be obtained with a Lorentzian distribution in critical temperatures about a nominal $T_c$ of 150°K with a half-width of about 2°K.

REFERENCES

1. WICKMAN, H.H. and WERTHEIM, G.K., *Spin Relaxation in Solids*, in *Chemical Applications of Mössbauer Spectroscopy* (Academic Press, New York, 1968).

2. LEVINSON, L.M., LUBAN, M. and STRIKMAN, S., *Phys. Rev. 177*, 864 (1969).

3. KHURGIN, B. et al., *J. Phys. Chem. Solid 31*, 49 (1970).

SESSION 7

HYPERFINE INTERACTION
IN
STRIPPED ATOMS, I

Chairman:

S.S. HANNA

*Stanford University*

# RECOIL INTO VACUUM*

ROLF NORDHAGEN

*Physics Institute, University of Oslo, Norway*


1. INTRODUCTION

When nuclear products recoil into vacuum, one can observe a striking phenomenon of strongly perturbed angular correlations. This has, in recent years, led to considerable interest in understanding the hyperfine interaction in highly ionized, free atoms. In the following, recent work pertaining to the recoil into vacuum will be discussed, while the closely connected subject of recoil into gas will be covered in the review by G. Sprouse [1].

Already the early works on perturbed angular correlations (Hamilton, Goerzel [2]) realized the importance of a hyperfine interaction between the nuclear magnetic and electric moments and the surrounding field from the electron structure. The first work in which the effect was observed in vacuum was by Flamm and Asaro [3]. Here, the perturbed angular correlations from nuclei recoiling into vacuum was measured following α-decay. Unfortunately, until heavy ion beams from Tandem and Linear accelerators became available, the difficulty in preparing nuclear decay products in vacuum with high recoil velocities made further investigations difficult. However, with present accelerators an extensive effort pioneered by the Weizmann Institute group [4] has been initiated for studying perturbed angular

---

*Invited paper.

correlations in nuclei recoiling into vacuum and gas. This and other work, cited in the following, has thrown considerable light on the subject, but some of the basic atomic processes involved are still poorly understood.

Primarily, three aspects of the process have been of interest:

1) Understanding the hyperfine interaction in highly excited free atoms in vacuum.
2) Using the interaction as a tool in measuring the magnetic moment of excited nuclear states.
3) Utilizing the information on the interaction as an important correction to measurements where the alignment of nuclear products in vacuum must be known.

To the nuclear physicist, the last two aspects are of major interest. However, the main problem of understanding the atomic processes involved has directed much of the effort towards studying the hyperfine interaction itself.

## 2. THE HYPERFINE INTERACTION

### 2.1. *General Remarks*

The perturbation of nuclear alignment which is observed when atomic ions appear in vacuum can be understood as follows: The electron configuration gives rise to a resultant electron spin J, and this spin, together with the nuclear spin I, will add to an atomic spin F. Through the magnetic and electric fields associated with the electron configurations and nuclear moments, and in the absence of an external field, I is coupled to J with a precession frequency $\omega$ (the hyperfine interaction). In vacuum, F stays fixed in space, and due to the hyperfine coupling the orientation of I changes relative to F and thus also in space.

The coupling is predominantly magnetic. For instance, with electrons in S-orbits, the magnetic dipole-dipole interaction in the free atom is large relative to the electric quadrupole interaction. Thus the interaction is dominated by a magnetic hyperfine structure constant

$$a = \frac{\mu}{\hbar} \frac{H_J}{IJ} = g B \frac{\mu_N}{\hbar} \qquad (1)$$

where, to simplify our notation we write the magnetic field from the electrons as $H_J/J = B$. The nuclear magneton is written as $\mu_N$ and $g$ is the nuclear g-factor.

The re-orientation of I in space is experimentally observed as a change in the substate population for the initial nuclear alignment. The most common method for studying the re-orientation is by angular correlation measurements.* We use the common notation for the angular correlation function [5]

$$W(t,\theta) = \sum_k A_k G_k(t) \, P_k(\cos \theta) \qquad (2)$$

where the influence of the perturbing interaction on the intermediate nuclear state is contained in the time-dependent attenuation coefficient $G_k(t)$. In many cases the time-integrated effect of the perturbation is observed, giving a value for $G_k(\infty)$ which will be denoted by $G_k$ in the sequel.

## 2.2. Static Interactions

The case of a static free atom, that is where no changes in the value and direction of the electron spin J take place, has been carefully treated by Alder [6]. For the explicit assumption of a random orientation between I and J, $G_k(t)$ is obtained as

$$G_k(t) = \sum_{FF'} \frac{(2F+1)(2F'+1)}{(2J+1)} \begin{Bmatrix} F & F' & k \\ I & I & J \end{Bmatrix}^2 e^{-(i/\hbar)(E_F - E_{F'})t} \qquad (3)$$

and the time-integrated coefficient as

---

*Mössbauer-effect measurements have also been used to observe the substate population, e.g. by G.D. Sprouse, S.S. Hanna, and G.M. Kalvius, *Phys. Rev. Lett.* 23, 1014 (1969).

$$G_k = \sum_F \frac{(2F+1)^2}{2J+1} \begin{Bmatrix} F & F & k \\ I & I & J \end{Bmatrix}^2 + \sum_{F \neq F'} \frac{(2F+1)(2F'+1)\begin{Bmatrix} F & F' & k \\ I & I & J \end{Bmatrix}^2}{(2J+1)(1+(\omega_{FF'}\tau)^2)} \qquad (4)$$

for an intermediate nuclear state with mean life τ. We can illustrate the form of this equation by taking the special case of J = 1/2 [6]

$$G_k(J = 1/2) = 1 - \frac{k(k+1)}{(2I+1)^2} \frac{(\omega\tau)^2}{1+(\omega\tau)^2} \qquad (5)$$

The precession frequency ω for a purely magnetic I-J coupling is given by

$$\omega_{FF'} = \frac{1}{2} a \, (F(F+1) - F'(F'+1)) \qquad (6)$$

As in other static hyperfine interactions with random initial orientation of the interacting spins, a 'hard core' value exists for $G_k$:

$$G_k = \sum_F \frac{(2F+1)^2}{2J+1} \begin{Bmatrix} F & F & k \\ I & I & J \end{Bmatrix}^2 \qquad (7)$$

This is the value obtained for $G_k$ in cases of very long nuclear lifetimes and is independent of the interaction strength.

## 2.3. *Time-Dependent Interactions*

In most observed cases, the angular-correlation coefficients are found to be attenuated well below the hard-core value. This is indicative of a time-variable interaction, which has been treated by Abragam and Pound [7]. In this case

$$G_k(t) = e^{-\lambda_k t} \qquad (8)$$

and the time-integrated coefficient becomes

$$G_k = (1 + \lambda_k \tau)^{-1} \qquad (9)$$

where we express the relaxation constant as

$$\lambda_k = p_k \omega^2 \tau_c \tag{10}$$

with the numerical constants $p_k$ being $p_2 = 2$, and $p_4 = 20/3$ for the magnetic case. The effect is similar to the interaction in liquids, where rapid changes in the magnitude and direction of the hyperfine field cause a time-varying precession of the nuclear spin. The correlation time $\tau_c$ is the time taken for a significant change in the varying field. It is seen that for large values of $\lambda_k \tau$ the angular correlation becomes isotropic.

In vacuum, the only source of variation is changes in the magnitude and direction of the electron spin J and the field associated with the electron configuration. These variations are due to changes in the electron structure of the excited, free atom, mainly caused by Auger electrons and optical transitions. With highly stripped ions (few electrons) optical transitions dominate. It is important to note that as the correlation time becomes short, the attenuation disappears. This is generally the case for relaxation in metals, when $\tau_c$ is too short to permit I to 'follow.' The same applies to relaxation by collisions in gas, which 'restores' the angular correlation with pressure.

To obtain a significant value for $G_k$, we have to fulfill the condition

$$\lambda_k \sim 1/\tau \tag{11a}$$

the relaxation constant roughly equal in magnitude to the nuclear decay constant. For the Abragam and Pound theory [7] to be valid, the time-variation must be caused by a large number of elementary field changes. This implies the condition

$$\omega \tau_c \ll 1 \tag{11b}$$

and consequently

$$\omega \tau \gg 1 \tag{11c}$$

Taking (somewhat arbitrarily) $\omega\tau_c < 10$, we can estimate the average field strength necessary to obtain an observable value of, e.g., $\lambda_2$. If we ignore spin dependent effects, $\omega$ is given in the time-dependent case by

$$\omega = (\mu_N/\hbar)gB \tag{12}$$

Taking low-lying excited states in rare earths as an example for the nuclear lifetime,

$$10^{-11} < \tau < 10^{-8} \text{s} \tag{13a}$$

and taking g as the 'rotational' value for even-even nuclei,

$$g_R \simeq 0.4$$

we obtain for B

$$250 \text{ MG} > B > 250 \text{ kG} \tag{13b}$$

This is the range of values we expect for the very large fields from unpaired electron structures in highly excited atoms, in particular from s-orbits.

The above simple formalism depends critically on the correlation time $\tau_c$. The derivation [7] is performed under the explicit assumption of a correlation time short compared to the nuclear lifetime. The time $\tau_c$ depends on the atomic processes involved, that is on how fast J is changing. As this process takes place in highly excited atoms, very little knowledge is available on this vital parameter. A further consequence of this lack of insight into the process is that the validity of the condition 11b above is not too certain. However, rough measurements of $\tau_c$ exist, and will be referred to later. Also, Blume [8] has derived expressions which can be applied to cases with different relative values of the correlation time and the nuclear mean life.

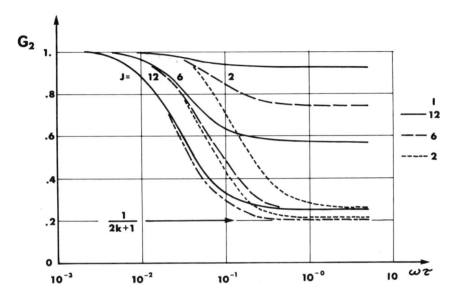

Fig. 1. The perturbation coefficient $G_2$ for different values of nuclear spin I and electron spin J as a function of $\omega\tau$ in a static, free atom. Calculation based on formula (4) by Alder [6].

## 2.4. *Spin Dependence*

In the static case, the interaction is strongly spin dependent. It has recently been found [9] that this also applies to the time-dependent case, where the relaxation constant $\lambda_k$ is found to be a function of the relative magnitudes of I and J. When I is large and J is small, the resultant F is highly aligned with I and the precession of I around F will be small (the static precession). This is clearly brought out if we plot Alders [6] formula for $G_2$, formula (4), on Fig. 1, as a function of $\omega\tau$ for different values of I and J. As can be seen, $G_2$ approaches the unperturbed value 1.0 very rapidly for high values of I and small values of J. For large J and I small, the static hard-core limit

$$G_k = (2k+1)^{-1} \qquad (14)$$

is reached for large values of $\omega\tau$.

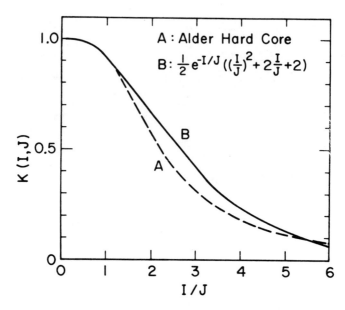

Fig. 2. The spin-dependent reduction factor K(I,J) in the relaxation constant for the time-dependent perturbation (section 2.4).

In addition, changes in small values of J will for large values of I give rise to small perturbations, that is for the time-dependent perturbation itself. To evaluate this, a knowledge of the time-variation of the J distribution is necessary, as well as that of the associated magnetic field. A rough way of introducing the spin dependence into the relaxation constant is to write as in [9]

$$\lambda_k = p_k \omega^2 K_k(I,J) \tau_c \qquad (15)$$

with $K_k(I,J)$ as a spin-dependent reduction factor. For large J relative to I, $K_k$ is evidently 1.0. From the data on Fig. 1 we can estimate a dependence of $K_2(I,J)$ on I and J, and we find the function given as A on Fig. 2. This function is, to a good approximation, dependent only on the ratio I/J. Simple time-dependent models have also been tried [9], giving either the function B in Fig. 2, or a relationship $K_2(I,J) = \exp(-2\ I/J)$. In both cases one assumes that the electron-spin distribution is exponential with J as the mean value. In the case of curve B (Fig. 2) the correlation time is taken as short relative to the time taken to run through the J distribution, in the other case these times are taken as comparable. In both cases the times are considered short relative to the nuclear lifetime. However, the relations between these different times are by no means clear, and the understanding of these processes is far from satisfactory. A knowledge of the J-distribution in highly excited free atoms is evidently desirable.

## 3. EXPERIMENTAL DISCUSSION

### 3.1. *Nuclear Alignment*

Heavy-ion beams have considerably widened the potential usage of the 'recoil into vacuum' phenomenon. Nuclear reactions are a well known source of nuclear alignment, and in particular a major effort has been directed to the study of the Coulomb-excitation reaction. Taking this reaction as an example, we note the well known fact that high nuclear alignment is achieved by observing the reaction products in coincidence with beam particles scattered along the beam axis. With an even-even target nucleus, the recoiling Coulomb excited products will be populated almost exclusively in their m = 0 magnetic substates. (Slight deviations from m = 0 are due to the finite solid angle of the particle counter, and to any de-orientation effects in the target foil.) Ideally, $A_2$ and $A_4$ coefficients as large as 0.71 and -1.72, respectively, can be achieved for the direct excitation of $I = 2^+$ first excited states.

The attenuation of this alignment is observed as the product nuclei recoil into vacuum, normally as the time-integrated coefficients $G_k$. In a few recent cases [10,11] the actual time dependence is also measured, as in the contribution by the Wisconsin group (Polga et al. [10]) shown on Fig. 3. Here, γ-ray angular correlation from the Coulomb excitation of $^{150}$Sm is observed as a

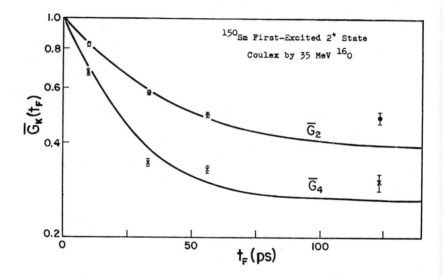

Fig. 3. The time-dependent attenuation coefficients for the angular correlation of γ-rays following Coulomb excitation of $^{150}$Sm, observed in vacuum and as a function of stopper distance in a 'plunger' type recoil-distance method experiment by Polga et al. [10]. The distance is converted to the corresponding flight time. The residual alignment at large distances is due to decays in flight.

function of stopper distance in a 'plunger'-type recoil-distance method experiment.

Measurements of this kind may be able to distinguish between perturbations taking place as normally expected by many small interactions, or by a few violent ones (The Dillenburg-Maris effect [12]). Also, the theories of Blume [8] can be applied. Experimental evidence is indicative of deviations from the Abragam and Pound theory [7], as will be discussed by G. Sprouse [1]. However, statistical uncertainties are still too large to permit definite conclusions. Evidently, more accurate experiments are needed.

In addition to Coulomb-excitation, the following reactions have been used: The $^{12}C(^{12}C, \alpha)^{20}Ne$ reaction [13], observing decay γ-rays in coincidence with α-particles along the beam axis, again producing m = 0 substate populations, and heavy-ion (xn)-reactions [9] producing an m ∿ 0 alignment by the very large impact parameters and high angular momenta imparted to the recoiling products.

## 3.2. Electron Excitations

In the experiment by Flamm and Asaro [3], the production of atomic ions with strong electronic fields was achieved by letting products from α-emitting $^{243}Am$-nuclei recoil out of a thin source foil and into vacuum. The atomic excitations are produced either in the decay process itself or by inelastic collisions in the source deposit. The use of high-velocity heavy-ion beams has, in addition to the advantageous nuclear excitation possibilities, introduced a very effective method for producing recoils with velocities near several per cent the speed of light. Under these conditions the ions will be heavily stripped when moving in the target foil, and thus strongly excited. After penetration in matter of 5-10 μg/cm$^2$ [13] the ions will achieve equilibrium ionization, and as target foils are normally considerably thicker this ionization will be present upon leaving the target. The expected charge probability as a function of velocity for $^{20}Ne$ ions in zapon-foils (carbon) is shown on Fig. 4, taken from the work of Zaidins [14]. As can be seen, already at low velocities a large number of electrons are stripped off.

A recent Heidelberg experiment on the $^{12}C(^{12}C,\alpha)^{20}Ne$ reaction, exciting the $^{20}Ne$ first excited state at 1.63 MeV, (M.A. Faessler [13]) is particularly enlightening regarding the atomic excitation process. With Carbon beams from 17 to 35 MeV and α-scattering angles at 0 and 180°, values of v/c from 1.3 to 5.9% were obtained for the $^{20}Ne$ recoils. The observed value of $G_4$ is shown on Fig. 5, where $1 - G_4$ is plotted as a function of v/c. The pronounced rise in perturbation closely follows the rise in $\phi_9$ on Fig. 4. At the high velocities, the recoils exist as ions with charge 9$^+$. It is then

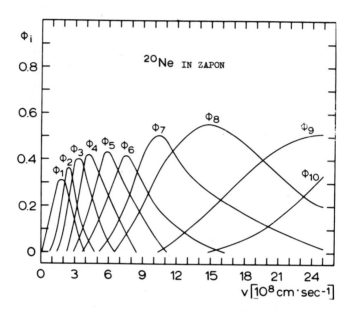

Fig. 4. The probability $\phi_i$ of finding $^{20}$Ne ions with i electrons stripped off, after penetrating zapon (carbon) foils. After Zaidins [14].

possible for Faessler to calculate the field expected from the hydrogen-like electron structure. For a single electron in the $1s_{\frac{1}{2}}$-state, the ionic ground state, a field of 167 MG is obtained. With the short nuclear lifetime (1.2 ps), and reasonable assumptions for the g-factor of the excited state, g = 0.5 - 0.7, this very large field - probably the largest ever encountered - explains the observed perturbation rather well. In this case the few existing electrons make it also possible to evaluate the influence of the most probable electron configurations, and only the ground state in the 9+ ion has a lifetime longer than $10^{-13}$s. Thus the interaction is static, and formula (5) is applicable.

In the more common cases, when heavier nuclei are recoiling at moderate velocities, only a fraction of the electrons are stripped off, and no detailed knowledge of the electron orbits is available. In earlier works [4]

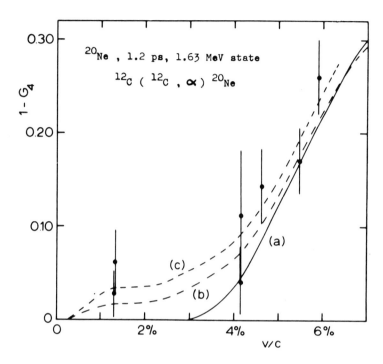

Fig. 5. The perturbation in vacuum of γ-ray angular correlations from the $^{20}$Ne first-excited state at 1.63 MeV, excited in the $^{12}$C($^{12}$C, α)$^{20}$Ne reaction, given as (1 - $G_4$) and as a function of v/c. From Faessler [13].

one hoped that the hyperfine field was a smooth function of Z and velocity. However, electron shell effects seem to be present. This is for instance seen in the recoil into vacuum work by a Rutgers group (J. de Boer et al. [15]) shown on Fig. 6, where $G_2$ and $G_4$ varies irregularly as a function of v/c, and thus with the number of stripped-off electrons. Similar evidence is found in a recent gas work in Rehovot [16].

3.4. *The Hyperfine Interactions*

The interaction has been shown to be predominantly magnetic. As an electric quadrupole interaction will

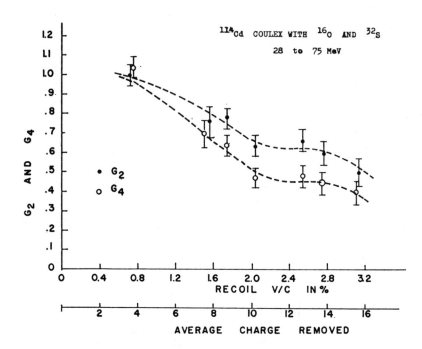

Fig. 6. The variation of the time-dependent attenuation coefficients as a function of recoil velocity, observed in vacuum following Coulomb excitation of $^{114}$Cd. Deviations from a $G_k = (1 + \text{const.} \ v^{0.1})^{-1}$ dependence taken as evidence for electron shell effects. From J. de Boer et al. [15].

give rise to a ratio $G_2/G_4$ different from the magnetic case ($p_2 = 34/5$, $p_4 = 4$, $\omega = (eQ/8\hbar)(d^2V/dz^2)$ for an $I^\pi = 2^+$ state), a plot of $G_2$ versus $G_4$ has been prepared by the Rehovot group for a number of cases (Fig. 7), and all support the magnetic assumption. As one expects the hyperfine interaction during relaxation in gas to be identical to the one in vacuum, a recent measurement for $^{172}$Yb recoils in gas [16] can be taken as additional strong evidence for a purely magnetic interaction. In this case, the nuclear quadrupole moment is particularly large. Even then, $G_2$ and $G_4$ show excellent agreement with the magnetic values. Similar evidence is found by Ansaldo and Grodzins [17] for trans-lead nuclei (U, Ra, Rn) at low velocities.

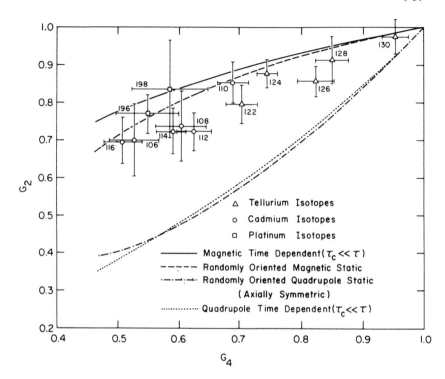

Fig. 7. The attenuation coefficients $G_2$ and $G_4$ for a number of measured correlations, indicating agreement with values expected for a magnetic hyperfine interaction. From Ben-Zvi et al [4].

The $G_2/G_4$ ratio can also be used to distinguish between other aspects of the interaction, e.g. the theories of Dillenburg and Maris [12] or Blume [8]. In a recent study by Brenn et al. [18] evidence is also found in these cases for a deviation from standard Abragam and Pound theory [7]. However, here also the statistical uncertainties are too large to permit definite conclusions.

One of the assumptions of the theory is that the initial electron spin is randomly distributed relative to the nuclear alignment. Similarly to the strong nuclear alignment in large-distance impact reactions, one might

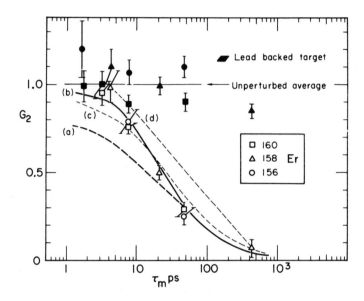

Fig. 8. The attenuation coefficient $G_2$ observed for a number of γ-ray angular distributions from excited states in the nuclei $^{156,158,160}$Er, excited in the Sn(HI, xn)Er reaction and recoiling into vacuum [9]. The short-lived states have spins up to I = 10. Curve a, with a poor fit, indicates expected values for full perturbation of all states. Improved fits are obtained by assuming strongly reduced perturbations for high spins, curves b and c. Curve b is obtained for a spin-dependence given by function B (Fig.2) J ∿ 3, curve c by ignoring the perturbation for I ≳ 10-12. Curve d is the dependence obtained with small J values (∿1) in function B (Fig. 2).

fear that the electron configuration is similarly aligned. However, as pointed out by Faessler [13], the main contribution to the resultant magnetic field at the nucleus is not due to the electron orbits, but to the electron spins which are randomly oriented.

Except for the $^{20}$Ne measurement [13], all experiments in heavier and not nearly so completely stripped ions are found to follow the time-dependent theory. This is shown both by the time-dependent behaviour of $G_k(t)$ and by observing values of $G_k$ well below the hard-core limit. However, the actual value of the correlation time $\tau_c$ is difficult to establish. An additional handle on this parameter is obtained in recoil into gas experiments, and values near

$$\tau_c \sim 3 \times 10^{-12} s$$

are found by the Rehovot group [4]. In practice only the product $\omega^2 \tau_c$ can be found in recoil into vacuum experiments. Most quoted field strengths from such experiments are obtained with the $\tau_c$ value given above. In this way, values between 20 and 40 MG are inferred for the fields encountered in heavier nuclear recoils.

Also, the spin dependence (formula (15)) complicates the time-dependent cases. Fig. 8 shows the results of the heavy-ion (xn)-reaction experiment in Berkeley, where a number of states in even-even Er-nuclei were excited simultaneously, and thus the influence of the interaction on states with widely different spins (and lifetimes) could be compared directly. As mentioned before (see Fig. 2) strongly reduced perturbations are observed at high spins, due to the reduced effectiveness of a smaller electron spin J to sway the nuclear spin I.

Also, the spin dependence may have important consequences for recoil into gas cases. Fig. 9 shows a summary of some Rehovot [16] anisotropy measurements as a function of gas pressure. The overall behaviour of this type of dependence will be discussed in the review by G. Sprouse [1]. Here, the possibility of explaining the low-pressure behaviour as a result of an electron-spin increase will be mentioned. As the collisions in gas start to occur, we may expect a large transfer of J values to the recoiling ions. When this J-transfer leads to an increase in the average value of the J-distribution, an increase in K(I,J) from the vacuum

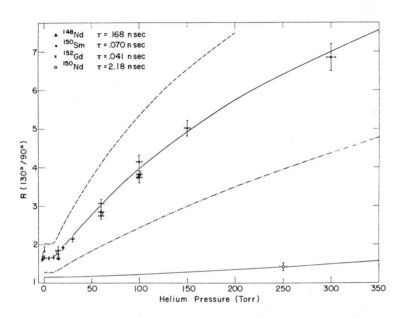

Fig. 9. Gamma-ray anisotropies measured as a function of gas pressure. From the Coulomb excitation of $2^+$ excited states in rare-earth nuclei, reference [16].

value will occur. For low pressures this increase may compensate for the decreased perturbation due to the decrease in correlation time with pressure.

### 3.5. *Experimental Corrections and g-Factors*

With this considerable amount of uncertainty in the parameters, how can the information be used to extract g-factors and to perform experimental corrections? Evidently, when values of

$$\omega^2 \tau_c = \text{const.} \; H^2 g^2 \tau_c$$

are obtained from measured $G_k$ coefficients, values of g in nuclei
    a) of the same ionic species,
    b) for excited states with identical spins, and
    c) at the same recoil velocities,

can be directly compared, as the const. $H^2\tau_c$ should be the same in all cases. To obtain comparison at different values of Z and v, interpolation is normally used. As discussed by the Rehovot group [16] most measurements are consistent with a velocity dependence

$$\underset{\sim}{H} \propto v^{0.6}$$

The electron shell effects found in the same work and by the others [15] limit the validity of the interpolation procedure. Over small regions of v, however, this method should be applicable. A dependence of H on Z is also observed, although small. In principle also here shell effects should be present. Again, interpolation over small ranges are performed. Normally, to extract meaningful data, pooling of larger volumes of information, both from vacuum and from gas measurements, is necessary [16]. In this manner, a considerable number of g-factors are actually evaluated, particularly in the rare-earth region. As an example, the only g-factor based solely on recoil into vacuum is the average g-factor obtained for the ground-band states of the three Er-nuclei with mass 156, 158 and 160 [9]. This value

$$\bar{g} = 0.37 \pm 0.03$$

was obtained by observing $G_k$-factors for a total of 8 correlations in the three nuclei, and finally comparing with the value of $H^2\tau_c$ obtained by extrapolation from a Sm-measurement with known g-factor.

To obtain the necessary corrections for other experiments where the angular distribution of reaction products emitted by recoil into vacuum must be known, usually separate experiments are performed to determine the values of $G_k$. In fact, many of the studies undertaken were initiated for this purpose. A typical case is the need for knowing the angular distribution of γ-rays in the recoil-distance half-life measurements.

To conclude, then, one is warned to proceed with caution when attempting to reach conclusions on nuclear and atomic properties based on recoil into vacuum experiments. On the other hand the present lack of detailed knowledge on the atomic processes involved should be regarded as a challenge. It is hoped that as more imaginative experiments are performed we will finally learn sufficiently about the interaction to reap the benefits of having very large magnetic fields at our disposal in the atom.

It may be fitting, when visiting a land of ancient Mediterranean culture, to recall the thoughts of a classical philosopher. Recoils into vacuum remind me, perhaps more than any other phenomenon in nature, of a well known part in the dialogue "The Republic" by Plato [19]. Here the human observers are imprisoned in a cave, chained to see only a blank wall. With a fire behind, the real world is seen as shadows. What we see in our perturbed angular correlations are indeed shadows on the wall from a bright world of violent interactions and excitations. And, as in Plato, the prisoners' enjoyment is to guess at the real world behind the shadows.

REFERENCES

1. SPROUSE, G., Recoil into gas, review at this conference.

2. HAMILTON, D.R., *Phys. Rev. 58*, 122 (1940).
   GOERZEL, G., *Phys. Rev. 70*, 897 (1946).

3. FLAMM, E., Ph. D. thesis, University of California, Lawrence Radiation Laboratory, report UCRL - 9325 (1960), unpublished.
   FLAMM, E. and ASARO, F., *Phys. Rev. 129*, 290 (1963).

4. BEN-ZVI, I., GILAD, P., GOLDBERG, M., GOLDRING, G., SCHWARTSCHILD, A., SPRINZAK, A. and VAGER, Z., *Nucl. Phys. A121*, 592 (1968).

5. STEFFEN, R.M. and FRAUENFELDER, H., in *Perturbed Angular Correlations*, E. Karlsson, E. Matthias and K. Siegbahn, eds. (North-Holland Publ. Co., Amsterdam, 1964), p.1.

6. ADLER, K., *Helv. Phys. Acta* 25, 235 (1952).

7. ABRAGAM, A. and POUND, R.V., *Phys. Rev.* 92, 943 (1953).

8. BLUME, M., in *Hyperfine Structure and Nuclear Radiations*, E. Matthias and D.A. Shirley, eds. (North-Holland Publ. Co., Amsterdam, 1968), p. 912.

9. NORDHAGEN, R., GOLDRING, G., DIAMOND, R.M., NAKAI, K. and STEPHENS, F.S., *Nucl. Phys.* A142, 577 (1970).

10. POLGA, T., RONEY, W.M., KUGEL, H.W. and BORCHERS, R.R., this conference.

11. GRAHAM, R.L., WARD, D. and GEIGER, J.S., Chalk River Nuclear Laboratories, CAP - APS and SFM meeting, June 1970.
    BRENN, R., LEHMANN, L. and SPEHL, H., this conference.
    ARMBRUSTER, R., DAR, J., GERBER, I. and VIVIEN, I.P, this conference.

12. DILLENBERG, D. and MARIS, Th.A., *Nucl. Phys.* 33, 208 (1962).

13. FAESSLER, M.A., Thesis, Max-Planck Inst. für Kernphysik, Heidelberg (1969/70), unpublished.
    FAESSLER, M., POVH, B. and SCHWALM, D., this conference.

14. ZAIDINS, C.S., Charge States of Heavy Ion Beams in Matter, California Institute of Technology Report (1962).

15. De BOER, J., ROGERS, I.D. and STEADMAN, S., this conference.

16. BEN-ZVI, I., GILAD, P., GOLDBERG, M.B., GOLDRING, G., SPEIDEL, K.H. and SPRINZAK, A., *Nucl. Phys.* A151, 401 (1970).

17. ANSALDO, E.J. and GRODZINS, this conference.

18. BRENN, R., LEHMANN, L. and SPEHL, H., submitted to
    *Nuclear Physics*.

19. PLATO, The Republic, Book 7, 514 ff (trans. H.D.P.
    Lee, Penguin Classics, 1955) p. 278.

DISCUSSION

G. GOLDRING:  I would like to make a general remark.
It is clear that all the parameters pertaining to the
perturbation in vacuum· the hyperfine fields, the atomic
angular momentum and the correlation time depend on the
time as measured from the production of the atom. This
is also true in the statistical case, when the quantities
that are observed are averages over the whole atomic
population. We refer to this time as the age of the
atomic population. For example, as you go to very long
times or very old age, clearly all ions will be in the
respective ground states and will give rise to a pattern
of perturbation very much different from that discussed
here.

It has been found as a phenomenological fact that
in the regions that have so far been covered reasonably
well by experiments, the total perturbation parameter,
that is the quantity $\omega^2 \tau_c$ is not strongly age dependent.
But one must also bear in mind that such a result is of
necessity of limited validity, both in accuracy and in
its range of application. Therefore all general conclusions, as distinct from answers to specific problems,
will be of a general and qualitative character. By
specific problems (which can be tackled to a large degree
independently of this inherent problem) I mean problems
like comparing g-factors of states with similar lifetimes, or average measurements like those presented here
of the Berkeley work. But if one refers to general
measurements like g-factor measurements, or details of the
atomic process such as the spin dependence, the quality
of these measurements is at present not good enough to
elucidate the details of the process.

I would like to add a point of a technical nature.

# RECOIL INTO VACUUM

In many if not most of these measurements one works in the convenient set-up of back scattered ions which give a very distinct and clearly analyzable angular distribution pattern. One usually works under the assumption that this is adequately represented as a reaction product being scattered back at 180°. Now, for very simple technical reasons, the counter which observes the backscattered particles has a finite angle, and usually this is assumed to incur only a small modification of the process. In Coulomb excitation this is certainly correct, but in reactions in light nuclei this is not necessarily always true, and I would like to show here a slide to illustrate this point [Fig. A].

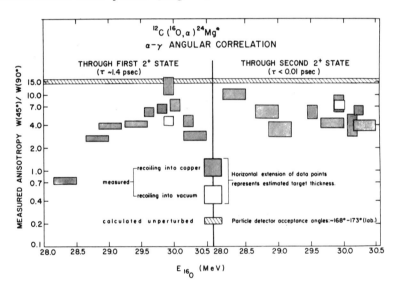

Fig. A

This refers to the $^{12}C + ^{16}O$ reaction, which is very similar to the $^{12}C + ^{12}C$ reaction used by Faessler et al. Here one expects from pure back-scattering a ratio of about 15 to 1 between gamma rays at 45° and at 90°. Actually, with a particle detector of an average opening angle of about 170° one observes angular distributions very different from the expected one. Clearly the process here is such that the angular distribution of the gamma rays depends violently on the angle of the emerging alpha. It is important to bear in mind that in such reactions in light nuclei one has to investigate the process much more closely and in detail before one can reach any exact conclusions about the perturbation in vacuum or any other quantity depending on the angular distribution of the gamma rays.

R. NORDHAGEN: I think that it is worth noting that you are here operating in what is called the region of Ericson fluctuations. These are the reactions which were presented some years ago as nuclear molecules, particularly because you observe extremely large angular momentum transfers. The distinctive difference between this reaction and the Coulomb excitation reaction is that in Coulomb excitation you have a smoothly varying yield, but here you get fluctuations which look like huge resonance structures. Evidently here you have a particularly nasty case of these fluctuations, whereas in Faessler's work, which we have been scrutinizing very carefully in view of this information, there do not seem to be any indications of these terrible things which you see here, which just again shows you that the life of the experimenter is not easy, and should not be so.

S.S. HANNA: I would just like to add that in his work Faessler actually checked each point, i.e., checked the angular distribution, presumably by recoil into a substance like gold, and therefore his points are presumable O.K.

D.E. MURNICK: I would like to get back to the atomic physics for a moment. I agree with what you have said and with what Prof. Goldring added. However, I do not think it is true that the recoiling atoms within the nuclear lifetime necessarily reach the atomic ground state. There are some well known metastable states, in particular for the 1 electron atoms such as neon $9^+$ there is the 2s metastable state, which lives for about $10^{-8}$ seconds. There are also well known metastable states for the 2 electron atoms, one of which was studied by the Oak Ridge group recently. Also, in particular for the neon experiment which was reported, only 40% of the

recoiling ions are in the 9+ state, at most. How then
could the experiment be analyzed in terms of this 160
MGauss field, where a certain fraction of the recoils
are in 10+, 8+ and 7+ charge states?

R. NORDHAGEN: Unfortunately I do not have Faessler's
thesis here. However, I know that he calculates all the
electronic configurations which are possible, and he then
shows that of the possible orbits he excites all of them
have lifetimes less than $10^{-13}$ seconds.

D.E. MURNICK: That is certainly not true for the $2s_{\frac{1}{2}}$
state for the 9+ atom.

R. NORDHAGEN: Yes, but his curves B and C take these
other states into account.

P. KIENLE: It is quite clear that the hyperfine interaction in a 2s state is much smaller than in the 1s state,
so even if you take it into account it does not make much
difference.

S.S. HANNA: I think we are some way off from actually
using calculated values of hyperfine fields to compute
g-factors. That is why one uses the gas method, I think,
which may be only something you do in a transition period
until you learn enough about the atomic structure.

H. SPEHL: I would just like to point out that nearly all
of the conclusions so far drawn from the recoil into
vacuum experiments are from time integral measurements.
But I think there is only one way to obtain reliable
information, namely by time-differential measurements.
We have already heard that there are many problems involved. It seems to me that there are really some hints
that there is a deviation from the Abragam and Pound
theory. It may even be more complicated, as has been
mentioned; perhaps after some time the interaction
becomes static, and so on. If you do time integrated
experiments you will never observe this, of course. You
have to rely, for the near future, only on time differential work.

L. GRODZINS: I have a number of questions and remarks that are all along the same lines as Prof. Goldring's. The first is just a note to correct a remark that you made. The highest magnetic hyperfine fields that have actually been observed are $10^{15}$ Gauss, and fields observed in the PAC are not even close to that. $10^{15}$ Gauss is the field due to a 1s muon at a nucleus which splits a nuclear level.

The first remark is about the Faessler work. It is in some sense unfair to compare the charge distribution after a reaction to the charge distribution observed after a particle has gone through its Zapon film; a particle which goes through its film has an equilibrium charge distribution, and in a reaction, provided you have a thick enough target and things happen close to the upstream surface, then presumably something like an equilibrium distribution will also pertain. On the other hand, the reaction itself is violent and probably kicks out deep inner electrons, and therefore you can use the Zapon data as a guide only, and perhaps that is all that Faessler used it for.

The second point is a point that was made in one of the papers of Ben-Zvi et al., namely that there is a large amount of angular momentum in the system following the violent collisions while passing through matter. If you try to calculate transition rates, the chances of your coming even close to reality, even in the lightest atom, are probably very small indeed.

Finally, it was already pointed out in the Bohr and Lindhard paper on charge exchange that you can get Auger processes after the particles have left the foils. The ions coming out of the foil can have several electrons which are highly excited so you get Auger effect taking place, a very violent deexcitation. Nobody really knows how to calculate the lifetimes of these transitions because again we are dealing with a very perturbed situation. Hans Betz and I have recently pointed out that the process may be more important than was first thought; whether that is correct or not is not known; there is no experimental evidence yet for this. But the fact is that one may very well have to take large amounts of Auger effect into account, and these should have lifetimes of the order of $10^{-15}$ sec.

Finally, I have a question to ask. Could someone explain to me how optical transitions have lifetimes which are much shorter than $3 \times 10^{-12}$ sec, which I believe you have to have in order to have a fairly strong spin axis change in a time of $3 \times 10^{-12}$ sec?

G. GOLDRING: The Auger electrons which you mentioned will speed up the decay very considerably.

L. GRODZINS: Well, if the Auger electrons speed up the transitions, then one should think that one should see very strong shell effects. I do not see how you can get around the fact that the high magnetic hyperfine fields are due mainly to unpaired s electrons. There may be orbital contributions but by and large one would think that most of the large field is due to unpaired s electrons. Consider xenon, for example; the Auger process should be over very quickly. After that point is reached, then any time dependent interaction should by and large not involve the s electrons which are well shielded by some 10 to 14 electrons. How do we get around this problem?

R. NORDHAGEN: I might comment on that. I think that we are facing a situation here where you have a very complex type of electron structure in detail, and that is just what you want - I mean, you want so many things to happen that the whole process more or less averages out. And I think that, first of all, at the present stage you can not hope to distinguish experimentally between the different effects which you are talking about. It depends, of course, whether you are definitely interested in the atomic processes as such, or just interested in utilizing the effect, say for g-factor measurements. But it is quite clear that we are absorbing in $\tau_c$ a multitude of sins, and we do not know anything about it yet. Of course it is very interesting to speculate on what might go on in there. But I do not think that as yet we have any hope of sorting out the different processes.

The remark was made the other day that if we could somehow find experiments which accentuate one aspect or the other, we might finally find a way to sort things out. But at the present stage, we are far from this. Also personally I am of the opinion that one here has to play the game just as well as one can, and see what one can get out of it, and if we start to be too finicky about details we just will not get anywhere. In that way, I think, the Faessler experiment is very important, because the surprising thing about this experiment is that it shows some resemblance to what we think is going on. It is an illustration that the general ideas of the process are more or less in the right ball park, but there can be finer details to it, of course, which we do not see. I think it is very surprising to see this curve rising exactly at the same velocity as where the $9^+$ ions start to appear. That is the main point of the Faessler experiment. And there seems also to be some numerical resemblance to what is actually going on. I think that is very nice, because we seem to be on the right track even if we do not understand everything perfectly.

# HYPERFINE INTERACTIONS IN TRANS-LEAD ELEMENTS VIA $\alpha$-$\gamma$ PAC*

E.J. ANSALDO and L. GRODZINS

*Laboratory for Nuclear Science, Massachusetts Institute of Technology*

(Presented by L. Grodzins)

ABSTRACT

A systematic study of the hyperfine interactions acting on trans-lead elements after alpha decay recoil implantation is being pursued in a variety of environments. The interactions in copper are of quadrupolar nature, time-fluctuating for the actinides, while the interactions in vacuum are magnetic. The effective magnetic field in ferromagnetic metals has been measured in a few cases.

We report in this paper the results obtained so far in our continuing study of the hyperfine interactions acting on trans-lead elements after alpha decay-recoil implantation into copper, iron, nickel, and vacuum by means of DPAC and IPAC measurements [1,2,3].

The sources employed in this work consisted of the parent activities of $^{243-241}$Am, $^{238}$Pu, $^{232}$U, $^{228}$Th ($^{224}$Ra) and $^{226}$Ra evaporated on the metal foils or on mylar (for the vacuum experiments). The experimental apparatus was

---

*This work is supported in part through funds provided by the U.S. Atomic Energy Commission Contract No. AT(30-1)-2098.

a conventional fast-slow arrangement. Fig. 1 shows the configuration used for the time-integral measurements.

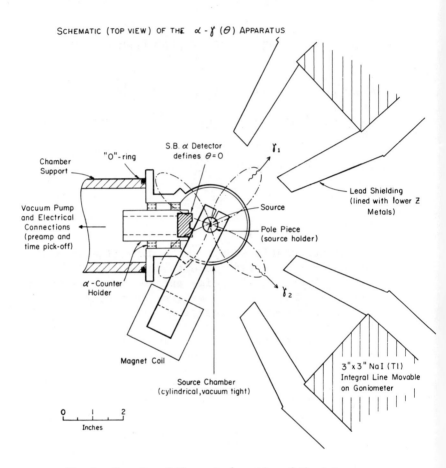

Fig. 1. Top view of the central portion of the α-γ apparatus. The open side of the source faces the alpha detector for the recoil into metals experiments, and away from the detector for the sources on mylar, to insure that the recoils end up in the selected environment.

The 3" × 3" integral line detectors were replaced by 1" × 1½" NaI(Tl) crystals coupled to fast photomultipliers and constant fraction timing devices for the time-differential measurements. The source activities were chosen, in all cases, so that the source thicknesses were small compared to the (calculated) range of the recoiling ions. The (possible) effect was tested in all cases by measuring the precession, in the ferromagnetic cases; after rotating the source around the symmetry axis in order for the recoils to travel through increased effective source thicknesses, and by finally washing off activity from the source surface and remeasuring. In such a fashion, an appreciable effect was found only for the $^{226}$Ra sources, not surprising given the comparatively long half-life (1602 y).

The integral attenuation factors obtained for the 2+ states of the even isotopes are shown in Fig. 2, along with the theoretical predictions for different interactions. The interactions in copper are clearly quadrupolar, in contrast to the magnetic interaction found for the rare earths implanted into copper [4]. A time differential measurement performed on $^{237}$Np in copper [2] has shown a non-static behavior of the interaction. The interaction in vacuum is of magnetic origin and probably time fluctuating. Those results will be discussed in greater detail elsewhere [3].

The interactions acting on the nuclei recoiling into iron and nickel have been shown to contain contributions from both an internal magnetic field (precession and attenuation) and a further attenuation which we attribute to a quadrupole interaction similar to the one found in copper [1,2]. Under such an assumption, the following magnetic fields have been determined: Rn (Fe) = +0.950 MOe [1], Ra (Fe) = -0.220 MOe [1], Th (Fe) = -0.230 MOe (preliminary result), and Np (Ni) = +0.170 MOe [2], with 20 to 30 percent accuracy.

In conclusion, it has been shown that the α-γ PAC is a very fruitful and easy technique for the investigation of the hyperfine interactions on the heavy elements. Considerable progress can be expected in such study from further research, both systematic and detailed, of the hyperfine interactions on the actinides using this and related

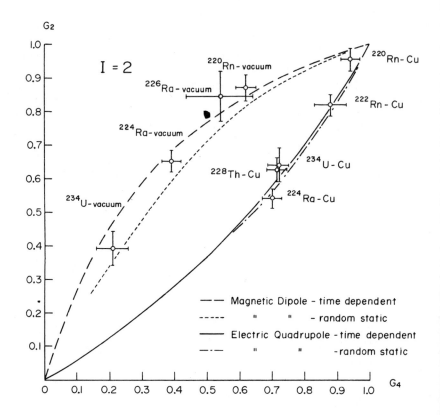

Fig. 2. Integral attenuation factors obtained for the 2+ states of the daughter nuclei shown. The Abragam-Pound theory has been used for theoretical curves in the time fluctuating cases.

techniques, $\gamma\gamma$ PAC, Coulomb excitation, isotope separator implantation and channeling.

REFERENCES

1. ANSALDO, E.J., GRODZINS, L. and KALISH, R., *Phys. Lett.* *30B*, 538 (1969).

2. ANSALDO, E.J. and GRODZINS, L., *Phys. Lett.*, in press.

3. ANSALDO, E.J. and GRODZINS, L., *BAPS*, *15*, 361 (1970), and to be published.

4. WADDINGTON, J.C., HAGEMANN, K.A., OGAZA, S., KISS, D., HERSKIND, B. and DEUTCH, P.I., *Proceedings Heidelberg Conference*, North Holland, 438 (1970).

DISCUSSION

B. DEUTCH: Just a one minute contribution. Shown on the figure is recoil for radon after α decay. The point I would like to make is that we are still in the infancy of solving the α recoil problem, due to source thickness and oxide problems that are present in the source-foil system. The particle acts, as was just mentioned, as a 100 keV isotope separator injector. Shown in the slide is a Lindhard-Scharff-Schiøtt range distribution curve of the radon after α recoil from a 8 µg/cm$^2$ source, through a 40 Å iron oxide layer compared with the Ra distribution from a source after a 50 keV isotope separator implant. In the latter one has a superposition of the two range profiles.

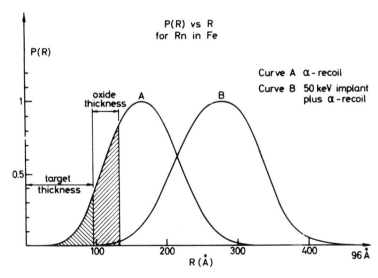

Fig. A.

We have recently measured (a preliminary measurement) the precession of these two cases, A and B. We find that implantation via B yields a 30% higher $\omega\tau$. We have fitted as a parameter the source thickness (the slide should show source thickness, not target thickness). It corresponds to only 8 $\mu g/cm^2$ thickness. To achieve a decrease of 30% in $\omega\tau$ due to the elimination of the precession from an oxide-source layer not only would there be a decrease of 30% in the measured precession angle but we would start picking up false quadrupole effects from the oxide film in the attenuation determinations. I really think this is still a problem with 100 keV implantation from the $\alpha$ particle recoil technique. Particularly since source thicknesses are probably greater than 8 $\mu g/cm^2$, and oxide layers probably increase after chemical source deposition.

L. GRODZINS: We did several things to try to obviate this problem. One was to turn the source so that the $\alpha$ particles came off at an angle in order that the source was effectively thicker and thicker. That, of course, is not an oxide layer, that is a source effect problem. The other thing that we did was to wash the source – just to take off layer after layer of the source material. The only time when we saw any effect under those circumstances was in the case of the long-lived $^{226}$Ra; the source was thick, and we corrected for that, but we did not correct for any oxide layer problem.

B. DEUTCH: Well, we did this measurement to learn how to do the alpha particle experiment, not to check the results of others. One cannot always implant the source – that is why one has to learn this method of $\alpha$ recoil. You do not often have such a convenient half-life for $\alpha$ measurements as in this radon case.

Another point I would like to make, since the uranium and plutonium hyperfine fields in iron were mentioned. I do not believe that the uranium field is necessarily in trouble, because it has actually been measured three times – once at Bell Labs., once at the Niels Bohr Institute, and once with a radioactive source via electron-$\gamma$ cascades at Aarhus. The fields always

# HYPERFINE INTERACTIONS VIA α-γ PAC

yield the same sign.

L. GRODZINS: I gave the wrong impression; let me correct it. I did not mean that the experiment was in trouble. We too have confirmed the sign. What I had in mind was the question of the interpretation. We already have trouble with understanding one negative sign (i.e. thorium) and we may now find ourselves having trouble with understanding a second one (uranium).

B. DEUTCH: The uranium and plutonium were both measured with electron-γ cascades. It is one of those very funny experiments where one can use the same correlation table, the same electron energy etc. The only exception is in the results. The rotation for the uranium case was opposite to that of plutonium. Hence, the relative sign of the two cases is different. There is another thing about these cases. Being an electron measurement, one can actually measure the stray field relative to the field of the iron. So you have one of those rare cases where you have an absolute calibration of the sign of the hyperfine field relative to the direction of magnetization in iron.

D.E. MURNICK: I would like to comment on the interpretation of the actinide fields, from a paper by Kaufmann and myself which will be presented at the Magnetism Conference next week. The fact that the field on uranium in iron is negative is not too surprising since the 5f electrons are not nearly as well shielded as the 4f electrons. In fact, several years ago Friedel suggested that there is a hybridization of the 5f's with the 6d band in solids, and this interpretation was used to explain the susceptibility of uranium-iron alloys and also the ferromagnetism of $UFe_2$. On this basis you would expect uranium in iron to act more like, say, platinum in iron, and to have a negative field, whereas for neptunium and plutonium the orbital contribution of the 5f electrons are important, and one gets a positive field as in the rare earth case.

L. GRODZINS: May I then ask - do you then expect a sign reversal again, i.e. a double crossing of the sign of the field?

D.E. MURNICK: Yes.

R.L. COHEN: I would like to make a further comment along these lines. I think it is important to realize that whereas in the rare earths the overwhelming part of the hyperfine interaction in most cases is from the 4f electrons, and only a small part comes from the contact term, in the actinide series the contact term is expected to be much larger, and the 5f term is much smaller than the 4f term. Depending on which half of the series you are in, these can either add or subtract. So the apparent discrepancy between the 4f and the 5f results is not really a physical problem.

I would like to point out that the temperature dependence of the hyperfine field measured in this way is enormously important and gives you a lot of information about whether this is a contact term or whether it is a 4f or 5f term.

H. DE WAARD: Experiments of Grodzins, Murnick and others indicate that the magnetic hyperfine fields in the series Th-U-Np-Pu etc. are shifted to more negative values than those found in the series Cs-Ba-La-Ce etc.

This behavior is consistent with the concept that the hyperfine field consists of the following contributions:
(a) conduction electron polarization giving a negative field increasing smoothly with Z (eg. $H_{CEP} \approx -300$ kG for Ag; $\approx -1$ MG for Au).
(b) core polarization increasing in the order 4f - 5f (and also 3d-4d-5d), also giving a *negative* contribution.
(c) overlap polarization predominant for high valency atoms and giving a *positive* field.
(d) orbital magnetism, only of importance in 4f and 5f shells.

In the region considered, the contributions (a) and (d) are predominant; (a) pulls the fields of the 5f

series to more negative values. Contribution (b) would
assist in pulling the fields to more negative values but
is probably smaller than (a).

F.C. ZAWISLAK: The first point I would like to make is
that the magnetic moment in radium probably was not
measured at Caltech. At least, I left Caltech half a
year ago and I do not know anything about it. The second
point is that we, in fact, measured $^{223}$Ra. First we
measured the magnetic moment of the state with a liquid
source, and then the internal field in iron, using a
dilute alloy of $^{227}$Ac in Fe. The measurement was made
by γ-angular correlation. We got for the g-factor +0.28
and for the field -100 kgauss. This field agrees in
sign with the result quoted by Prof. Grodzins, and if
you would use a g-factor of +0.3 for the 2+ state in
$^{224}$Ra it would agree also in value.

S.K. BHATTACHERJEE: I would like to clarify this matter
of $^{222}$Rn; the magnetic moment of the first excited state
was measured by the Uppsala group as 0.45.

J.D. BOWAN: I would like to make a comment concerning
the source thickness problem in α recoil work. At Bonn
we are in the preliminary stages of ion time of flight
experiments to study the slowing down process of α recoil
ions in gas. Of course in order to do this you have to
have very thin alpha sources in order to get a good time
of flight spectrum. For a $^{223}$Ra ($T_{\frac{1}{2}}$ = 12 days) source
evaporated on a 1 cm$^2$ mylar foil having 150,000 decays
per second, which is a reasonable activity for spin
rotation measurements, we find that the loss in ion
velocity in going from the source layer is less than 5%,
or an energy loss of less than 10 keV. By evaporating
one's sources on thin iron foils, where you could do
time of flight experiments, one could actually measure
the source thickness and the loss of ion energy.

# HYPERFINE INTERACTIONS OF FAST RECOIL NUCLEI IN GAS*†

G.D. SPROUSE

*State University of New York, Stony Brook, New York*

INTRODUCTION

The study of hyperfine interactions in recoil nuclei traveling with high velocities in vacuum or gas is a new field of interest that has been stimulated largely by the work of the Weizmann Institute group [1-3]. Many laboratories are now performing experiments related to this field [4-13]. The motivation for work in this field comes from several sources. The strength of the effective magnetic hyperfine interaction ($H_{eff} \approx 10^8$ Gauss) makes it the best tool available for measuring g-factors of states with $\tau < 10^{-11}$ sec, and one of the main goals of work in this field is to develop techniques for reliable, accurate g-factor measurements for short lived states. An understanding of the hyperfine interaction in vacuum is also necessary to make corrections to measurements that involve gamma ray angular distributions, such as lifetime measurements by the Doppler shift plunger technique [11], and reorientation effect measurements. The atomic information to be gained from studying these interactions is necessarily less detailed than the nuclear information because the interaction must certainly be an average over many different charge states and in each charge state, the interaction can be an average over

---

\*Invited paper
†Work supported in part by The National Science Foundation.

many different atomic configurations. However, by varying the velocity and the atomic number of the recoiling ion, some general aspects of the interaction can be ascertained.

## CHARGE STATES IN GAS

In order to put together a coherent picture of what happens when an excited recoil nucleus is knocked from a foil into a gas, it is important to first discuss the available information about how many electrons remain on the ion. Betz et al. [14] have shown empirically that the charge state distribution for ions moving in gas with $v > c/137$ and $Z \geq 7$, is approximately Gaussian. The mean number of charges removed from the ion, $\bar{\xi}$, and the full $e^{-1}$ width of the Gaussian distribution, $\Gamma$, are given by:

$$\bar{\xi} = Z\left(1 - \exp\left(-\frac{137}{Z^{2/3}}\frac{v}{c}\right)\right)$$

$$\Gamma^2 = 0.59 Z\left(1 - \exp\left(-\frac{5.8}{Z}\bar{\xi}\right)\right)$$

For most cases of interest, we can expand the exponential in the first formula and:

$$\bar{\xi} = 137 Z^{1/3} \frac{v}{c}$$

The equilibrium charge state distribution for silver ions with $v = 0.02 c$ is shown in Fig. 1. The distribution is rather broad and one would expect that this would tend to smooth out any large differences in the hyperfine interaction from adjacent charge states. However, there is some evidence for the charge state distributions to peak up near the more stable rare gas structures [15]. We will come back to this point later. When we are considering excited nuclei produced by Coulomb excitation with heavy ions, we can usually neglect any change in the velocity of the nucleus during the observation time because the rate of slowing down of the ion in gas is small. For example, a 20 MeV silver ion

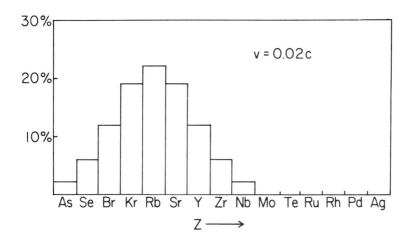

Fig. 1. Calculated equilibrium charge state distribution for silver ions recoiling in gas with v/c = 0.02. The horizontal axis is labeled with the chemical symbol of the element that has the same number of electrons as the silver ion.

recoiling into 1/2 atmosphere of He gas loses 5% of its velocity in 1 nsec. For lower energy recoils, the rate of slowing down is larger, and we have to take it into account in the analysis. We will restrict our discussion to states that decay while the ion has a large velocity. The situation where the ion stops and then the nucleus decays is much more complex [16, 17], and effects due to the chemical composition of the stopping gas have been observed. The chemical composition of the gas is also important for experiments with the low recoil velocities obtained from β-decay [18].

ANGULAR CORRELATION FOR NUCLEI RECOILING IN GAS

Fig. 2 shows the angular distribution of gamma radiation from $^{172}$Yb ($\tau_{1/2}$ = 1.6 ns) recoiling into 400 Torr of Argon gas at a velocity of 0.01 c. This ion has, on

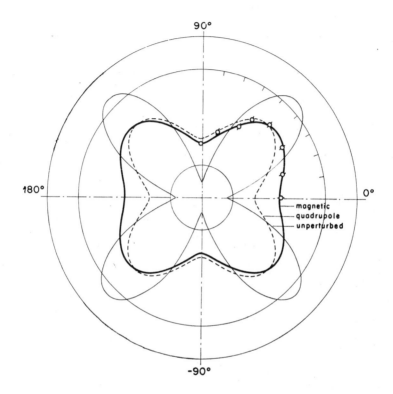

Fig. 2. Polar plot of the angular distribution of $^{172}$Yb recoiling into Argon. The perturbed curves are least squares fits for time dependent magnetic and quadrupole interactions using the theory of Abragam and Pound.

the average, the same number of electrons as Gadolinium. The angular distribution is fit very well by the Abragam-Pound theory with a time dependent magnetic interaction. Fig. 3 shows a plot of $G_2$ vs $G_4$ for this case and several other cases of recoil into gas. The Nd ions have, on the average, the same number of electrons as Cesium. The character of the interaction is apparently the same, even though in the case of Yb we are averaging over electronic structures near the middle of the rare earth region and in the case of Nd we are averaging over structures near

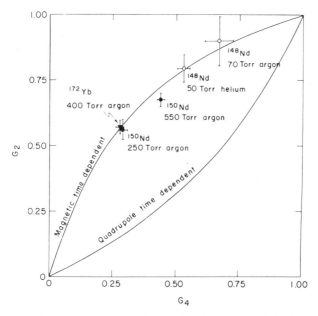

Fig. 3. The attenuation coefficients $G_2$ and $G_4$ for recoil into gas for different nuclei.

the limit of that region. It is now generally assumed for all cases that the predominant interaction is magnetic. This assumption should be experimentally verified for each case whenver possible.

The following assumptions have also been made in analyzing the early data for both recoil into vacuum and recoil into gas:

1) The free ion hyperfine interaction, $a\vec{I}\cdot\vec{J}$, can be replaced by a randomly oriented external field.

2) The hyperfine interaction is stochastic, i.e., it is static for a time, t, and then the direction of the effective field instantaneously jumps to a new direction. The average time

between successive jumps is $\tau_c$.

3) The nucleus precesses a small angle between collisions, i.e., $\omega\tau_c \ll 1$, and there are many collisions during the nuclear lifetime $\tau_c/\tau \ll 1$.

4) The average perturbation parameter, $\omega^2\tau_c$, does not change during the observation time.

5) The effective magnetic field varies smoothly with velocity and with the atomic number, Z.

The first assumption is just that the angular momentum of the electronic state J, is much larger than the angular momentum of the nucleus I. If this is true, then the quantum mechanical coupling of I and J can be replaced with a classical field. Qualitatively, since the angular momentum involved in the ion-atom collision is large, we expect the electronic shell to pick up a large amount of angular momentum after each collision. The measurements of the Berkeley group [2] indicate that for recoil into vacuum the average J is larger than $3\hbar$. For recoil into gas, we expect a larger average value of J because the ionic population is continually rejuvenated with high angular momenta and thus assumption 1) should be valid for states with $I \approx 2$ or less.

In order to analyze our results with the Abragam-Pound theory, we must show that the physical situation satisfies assumptions 2) and 3). The violent nature and short duration of the collision process makes it likely that the process is stochastic. For high gas pressures, $\tau_c \approx 10^{-13}$ sec, and a typical value of $\omega$ is $10^{11}$, so that the precession angle between collisions is small. However, the use of the Abragam-Pound theory to analyze vacuum data is not necessarily valid. For example, let us consider the $3/2^-$ level in $^{103}$Rh at 295 keV ($\tau_{1/2}$ = 7.2 psec). For $v = 5 \times 10^8$ cm/sec, the measured $G_2$ is 0.5. For $\tau_c(vac) = 3 \times 10^{-12}$ sec, we have $\omega\tau_c = 0.4$. Therefore, the conditions of the Abragam-Pound theory are not fulfilled. In addition, the $\omega$ that we used was the root mean square precession angle. For some atomic configurations, $\omega$ could much larger and we will be further from the range of validity of the theory. The value of $\tau_c$ (vacuum) of 3 psec is arrived at by comparison with calculated $\tau_c$ for gas, which depends on the poorly known atomic radii. One also expects the radius for a more violent collision that would reorient the atomic shell to be smaller than the tabulated values. This would tend to make $\tau_c$ larger than previously calculated.

Blume [19] has developed a theory for relaxation phenomenon in which it is not necessary to make the assumption that $\omega\tau_c \ll 1$. For the particularly simple case of a nucleus that sees a magnetic field fluctuating between the +z and -z direction, the Blume theory gives the following formula for $G_k(t)$:

$$G_k(t) = \frac{1}{2k+1} \sum_{N=-k}^{k} (\cos x\, Wt + \frac{1}{x} \sin x\, Wt) e^{-Wt}$$

where $x = (4\omega^2 \tau_c^2 N^2 - 1)^{\frac{1}{2}}$ and $W = \frac{1}{2}\tau_c$.

If integral correlations are observed, the average coefficient is given by:

$$G_k(\infty) = \frac{1}{2k+1} \sum_N \frac{1}{1 + \frac{\omega^2 \tau^2 N^2}{1 + \tau/\tau_c}}$$

In the limit of small $\omega\tau_c$ this formula reduces to the Abragam-Pound theory, and in the limit of $\tau_c \gg \tau$ (static interactions) it reduces to the formula of Matthias, Rosenblum and Shirley [19]. However, in the limit of $\tau > \tau_c$ and large $\omega^2 \tau \tau_c$, it has a hard core for a fluctuating interaction because the field flips along one axis.

To use the Blume simple model for the case we are considering where the effective field can point in any direction in space is an oversimplification and will certainly be incorrect in the region of the hard core. However, in the region where the number of fluctuations during the observation time is small and $\omega\tau_c \approx 1$, this model may give more physical results than the Abragam-Pound

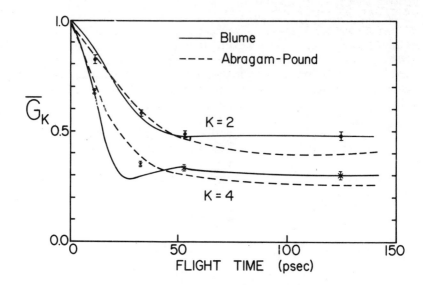

Fig. 4. The attenuation coefficients $G_2$ and $G_4$ for $^{150}$Sm recoiling into vacuum and stopping in a solid after a flight time $t_f$. The gamma rays are detected with NaI detectors, so that the nuclei that decay in flight are included. The parameters used for the Blume model were $\omega = 3.56 \times 10^{10}$, and $\tau_c = 20$ psec.

perturbation treatment. In other words, neither theory is strictly an accurate model for the vacuum case, but if the interaction is nearly static, i.e., $\tau/\tau_c \approx 1$, and if there is a large precession between jumps, $\omega\tau_c \approx 1$, then the Blume model is a better approximation than the usual Abragam-Pound treatment. As a test of these ideas, let us consider some time differential measurements of the Wisconsin group submitted to this conference [8]. Fig. 4 shows the attenuation factors $G_2$ and $G_4$ for $^{150}$Sm ($I^\pi = 2^+$, $E = 334$ keV, $\tau_{1/2} = 49$ psec) in vacuum as a function of the travel time to the Cu stopper. Notice how the experimental values for $G_2$ are constant after 50 psec and for $G_4$ are constant after 30 psec, while the Abragam-Pound theory continues to drop with increasing time. The Blume model reproduces the constancy but the shape at small times is not exactly correct.

# FAST RECOIL NUCLEI IN GAS

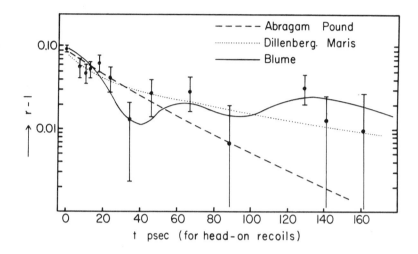

Fig. 5. The ratio r-1 for Yb recoiling into vacuum and stopping a a solid after flight time of $t_f$. The Abragam-Pound and Dillenberg-Maris theoretical lines are least squares fits to the ratio, while the curve labeled Blume is a typical curve for $G_2$ normalized at t = 0.

The Freiberg group [7] has done similar measurements with a somewhat different technique for natural Yb. In Fig. 5 the ordinate is roughly proportional to $G_2$ and the abcissa is time of flight. Again, the Blume model makes a better fit at long times, even matching the oscillations although they are not too significant statistically.

Another possible explanation of the flattening out of the perturbation factors at long times is that the Larmor precession frequency is not constant in time after the ion leaves the foil. If this is the case, the Blume theory for the simple case above, with the hard core, mocks the turning off of the perturbation at long times and fortuitously gives a good fit.

A more complex model of the interaction in which the field jumps to several different axes in space is soluble in the framework provided by Blume and should provide more realistic solutions, although they will not be in analytic form [20]. The recent theoretical work of Gabriel [21] may also be helpful in the present case.

TIME DEPENDENCE OF THE PERTURBATION

If we neglect the low pressure region for the moment, the Abragam-Pound theory should be valid for measurements with gas and we can see if there is evidence that the Larmor precession frequency of the fluctuating field is not constant in time while recoiling into gas. Fig. 6 shows $G_2$ vs gas pressure for the excited state of $^{107}$Ag recoiling into He gas with a maximum velocity of $8 \times 10^8$ cm/sec. Two different fits to the data are shown. The solid line is a fit with $\omega$ constant, and the dashed line is generated by the function:

$$G_2 = \frac{1}{1 + 2\omega^2 \tau \tau_c \, \exp(-2\tau/(n_0 \tau_c))}$$

This function corresponds to a model with the effective perturbation turned off faster in high density gas than in low density gas. The parameter $n_0$ corresponds to the number of collisions necessary to turn off the interaction. Although the difference in the two fits is not large, this declining field theory consistently gives a lower value of $\chi^2$ than the Blume theory or the Abragam-Pound theory.

The physical picture that Betz and Grodzins [22] have developed to explain the increase in charge state of ions leaving a foil may help to give the declining field theory some physical basis. The ion travelling in the solid achieves a high degree of excitation, and immediately after leaving the foil ($10^{-15}$ sec) loses several more charges by Auger emission. This higher charge state presumably gives a larger hyperfine field. As the ion travels in the gas, it picks up electrons and as the charge is neutralized, the field is made smaller. However, the number of collisions necessary to turn off the field derived from the fit to the gas pressure data is about 150, and this seems to be larger than the best

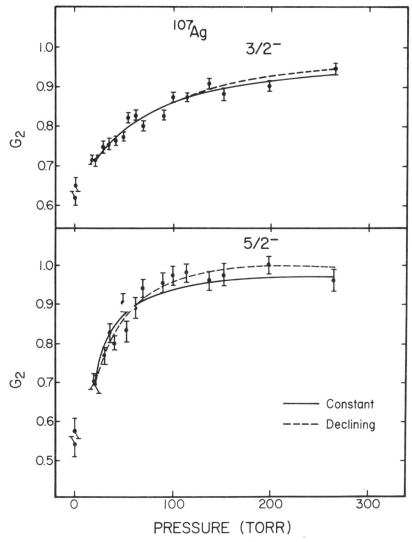

Fig. 6. Attenuation coefficient $G_2$ for excited states of $^{107}$Ag recoiling into gas, as a function of gas pressure. The solid line assumes $\omega$ constant; the dashed line is for a perturbation declining in time at a rate proportional to the collision frequency. For the $5/2^-$ state, the constant perturbation gives $\chi^2/f = 1.4$, the declining perturbation $\chi^2/f = 1.0$ (f = number of degrees of freedom).

estimates [14] of the number of collisions necessary to achieve charge equilibrium which is about 60.

TIME DIFFERENTIAL TECHNIQUES

Even if we do not completely understand the time dependence of the hyperfine interaction after the ion leaves the foil, g-factor ratios of states with different lifetimes can be measured precisely, provided that the interaction is averaged over the same time for both states. This can be done by introducing a stopper at a fixed distance from the target, and by making the flight time through the gas shorter than or equal to the lifetime of the shorter-lived state. If the perturbation is turned off completely in the solid, we can extract the r.m.s. Larmor frequency averaged over the flight time instead of averaged over the nuclear lifetime. Measurements of the r.m.s. average Larmor frequency for different averaging times for two states in $^{109}$Ag are shown in Fig. 7. For the $3/2^-$ state, the shortest flight time is comparable to the nuclear lifetime, so for this and all other flight times the averaging time is essentially the

TABLE 1. - Summary of g-Factor Measurements

| Nucleus | Energy (keV) | $\tau$ (psec) | $|g_{3/2}|/|g_{5/2}|$ | $|g|$ |
|---|---|---|---|---|
| $^{103}$Rh | 295.1 | 11 | $1.3 \pm 0.3$ | $0.6 \pm 0.15$ |
|  | 377.4 | 110 |  | $0.45 \pm 0.1$ |
| $^{107}$Ag | 324.6 | 8.5 | $1.25 \pm 0.3$ | $0.5 \pm 0.15$ |
|  | 422.6 | 43 |  | $0.4 \pm 0.1$ |
| $^{109}$Ag | 209 | 10 | $1.4 \pm 0.4$ | $0.55 \pm 0.15$ |
|  | 414 | 50 |  | $0.4 \pm 0.1$ |
| $^{104}$Ru | 357.7 | 75 |  | $\equiv 0.36$ |
| $^{110}$Pd | 373.8 | 58 |  |  |

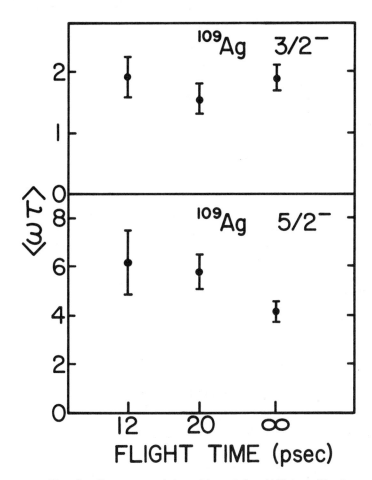

Fig. 7. The average interaction $\omega\tau$ for different fixed flight times for the lowest $3/2^-$ and $5/2^-$ states of $^{109}$Ag.

nuclear lifetime. For the longer lived $5/2^-$ state, however, the averaging time changes with the flight time and we see that the average values of $\omega$ decrease with increasing flight time, in agreement with a declining field model. Therefore, precise g-factor ratio measurements should be made with the same averaging time for both states.

Some recent measurements of g-factor ratios by the Stanford group are presented in Table 1. These ratios were obtained by a least squares fit of the Abragam-Pound theory to the pressure data above 20 Torr, and $\tau_c$(vacuum) was taken as 1.6 psec. Because the measurements on $^{103}$Rh and $^{107}$Ag were performed without a stopper, a 30% correction to these g-factor ratios was applied to compensate for the different averaging times for the two states.

Perturbation measurements of the $2^+$ states in the adjacent even-even nuclei were used to obtain an absolute calibration for the g-factors. The g-factors of $2^+$ states in the region fall around 0.36 and this value was used for the states in $^{104}$Ru and $^{110}$Pd. For the adjacent odd nuclei, the effective field was assumed to be the same as for these even nuclei. We can use the Z dependence of the effective field measured in heavier nuclei [3] to estimate that the error introduced by this procedure is less than 20%. The states used to determine the g-factor calibration have lifetimes of the same order as the $5/2^-$ states, but the $3/2^-$ states are considerably shorter. The 20% correction has been applied to the g-factors of the shorter lived states to account for the increased effective field at shorter times.

VELOCITY DEPENDENCE OF THE EFFECTIVE FIELD

Because the observed field is an average over many charge states, and in each charge state many electronic configurations can be present, the assumption that the field is a smooth function of Z and v is very plausible. The velocity dependence of the perturbation factors has been measured by several groups [6,9,10].

To measure the velocity dependence of the field after recoil into vacuum, we have used a technique in which data for a large range of velocities can be collected simultaneously. Fig. 8 shows a schematic view of the apparatus. We use a thick target (2.6 mg/cm$^2$) and the essential idea is that in the target the recoil nucleus is slowed down much more quickly than the projectile. The small slowing down of the beam of particles

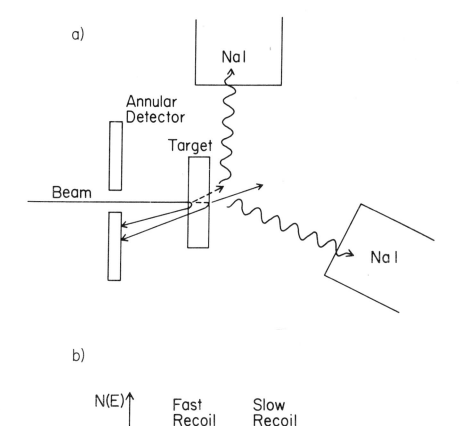

Fig. 8  a) Schematic view of the apparatus used to measure the velocity dependence of the attenuation coefficients. The target and detector dimensions are distorted for clarity.
b) Schematic representation of the particle spectrum observed in the annular detector.

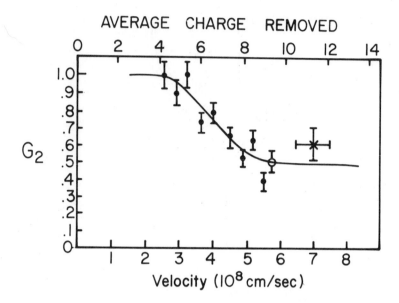

Fig. 9. The attenuation factor $G_2$ for the lowest $3/2^-$ state of $^{103}$Rh as a function of the velocity of the nucleus in vacuum. The open circle is from [11]. The (x) is from [5], and the error bar in the horizontal direction indicates that this measurement is done in singles and is an average over a range of velocities.

spreads the spectrum of the backscattered particles and can be used to locate the depth of the scattering in the target. By using the value of dE/dx tabulated by Northcliffe and Schilling [23] and the calculated recoil velocity, we can arrive at a one-to-one correspondence of the backscattered particle energy and the velocity of the ejected recoil nucleus.

The data were collected with a two dimensional analysis program written for a PDP-9 computer. The particle energy was used as the y coordinate and the x coordinate was divided into two halves for two NaI(Tl) counters at two different angles. A beam of 45 MeV oxygen ions from the Stony Brook FN Tandem Van de Graaff was used to excite the $^{103}$Rh target. The attenuation factor, $G_2$, for the lowest 3/2$^-$ state of $^{103}$Rh is shown in Fig. 9. There is a suggestion of a flattening out of the perturbation at $v = 5/10^8$ cm/sec. At this velocity an average of 8 charges are removed. A similar flattening out of the perturbation with velocity is suggested for $^{114}$Cd in [9], but at a higher velocity. However, the onset of the flat region in both cases corresponds to the same number of electrons on the ion. This correspondence suggests a way in which more precise experiments may be used to make relative g-factor measurements for nuclei of different Z.

If we consider two isoelectronic ions with different nuclear charges, in the lowest order approximation where we neglect the interelectron interactions, the radius of each electron will be smaller in the higher Z nucleus by an amount:

$$r_2 = r_1 \frac{Z_1}{Z_2}$$

The effective hyperfine field depends on $<r^{-3}>$, and we therefore expect for isoelectronic ions, that:

$$H_{eff}(Z_1,v_1) = \frac{Z_1^3}{Z_2^3} H_{eff}(Z_2,v_2)$$

where the condition for isoelectronic ions restricts the velocities to:

$$v_1 = v_2 \left(\frac{Z_2}{Z_1}\right)^{1/3} + \frac{(Z_1 - Z_2)}{Z_1^{1/3}} \frac{c}{137}$$

In order to make a comparison of the $^{114}$Cd and $^{103}$Rh g-factors, the perturbation parameter $\lambda = \omega^2\tau^2/(1 + \tau/\tau_c)$ of the Blume model was taken from the measurements of $G_2$, and is plotted in Fig. 10 for both cases. If we compare the regions of the flattening out for the two different nuclei, we see that the velocity condition for isoelectronic ions is satisfied. After we make the correction

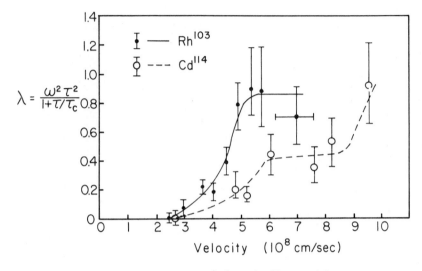

Fig. 10. The parameter $\lambda$ from the Blume model was plotted vs. velocity for both $^{114}$Cd and $^{103}$Rh.

for the contraction of the atomic shell due to the higher nuclear charge of Cd, we get:

$g(^{103}\text{Rh}, 298 \text{ keV}) = 2.3 \, g(^{114}\text{Cd}, 558 \text{ keV})$

We have used $\tau_c = 20$ psec for both cases since the dependence of the g factors on $\tau_c$ is weak. We can compare this result with g-factor measurements made in a more conventional way by Bhattacherjee et al. [24,25].

$g(^{103}\text{Rh}, 298 \text{ keV}) = (3.2 \pm 1.1) \, g(^{114}\text{Cd}, 558 \text{ keV})$

More stringent tests must be made to see if the scaling procedure outlined above is valid. In particular the velocity dependence of the effective field in tungsten would be a good case to measure because the Z dependence of the perturbation has been measured [3] for the even nuclei between $Z = 60$ and $Z = 76$.

## ORIGIN OF THE LARGE EFFECTIVE FIELDS

The magnitudes of the effective fields observed is generally too large to be explained by orbital contributions and the most promising explanation is that unpaired s electrons cause the large hyperfine interactions. Leisi [18] has used a central field approximation to enumerate the states of a many electron atom which are stable against Auger emission. For Iodine, many of these states have holes in the 4s shell, and could give effective fields of the right order of magnitude.

## EXPERIMENTAL TECHNIQUES

Although most measurements of attenuation factors have used the classic geometry of an annular detector at 180° in coincidence with a NaI detector, the measurements of the Stanford group were performed with a Ge(Li) detector in singles and very heavy ions as projectiles. The anisotropy observed was still large $(W(90°)/W(180°) \approx 1.4)$ and high velocity recoil nuclei were obtained. The disadvantages of this technique are that there is a distribution of recoil velocities, and the sensitivity to $G_4$ is very low because the singles angular distribution has practically no $P_4$ term.

A somewhat different technique utilizing the large Doppler shifts observed for nuclei excited by heavy ion projectiles is presented in [26]. Fig. 11 shows the gamma ray lineshape that would be observed in a Ge(Li) detector at 90° to the beam direction for E2 Coulomb excitation of states with different spins and different multipolarities of deexcitation radiation. The complex lineshape that is shown results from the fact that the gamma ray angular distribution is a function of the scattering angles of the projectile. In particular, for the 0-2-0 case, the valley in the middle of the lineshape arises because the radiation pattern for backscattered projectiles has no intensity at 90°. In addition, there is no radiation from nuclei excited by a scattering in the vertical plane through the beam direction. These two

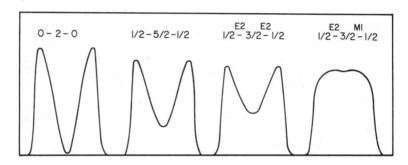

Fig. 11. Theoretical gamma ray lineshapes at 90° to the beam for four spin sequences following E2 Coulomb excitation. The maximum velocity of the recoil is v/c = 0.45, and a detector resolution of 3.5 keV has been folded into the lineshape for am 850 keV gamma ray.

situations are the only cases that give no component of velocity in the direction of the detector, and since they do not produce any radiation at 90°, there is zero intensity in the center of the line.

The depth of the valley in the middle is particularly sensitive to any influence which perturbs the very anisotropic angular distributions which come from the nuclei going directly forward. Fig. 12 shows how the lineshape changes as a magnetic perturbation is made stronger.

The extreme sensitivity of the valley height to small perturbation can be seen in Fig. 13. The ratio of the perturbed peak to valley ratio to the unperturbed peak to valley ratio has a steeper slope than $G_4$, and can possibly be useful in measuring small perturbations for very short lived states.

The lineshape technique may be useful for states where the excitation probability is too small to do a coincidence experiment, or where background gamma rays necessitate the use of a Ge(Li) detector.

# FAST RECOIL NUCLEI IN GAS

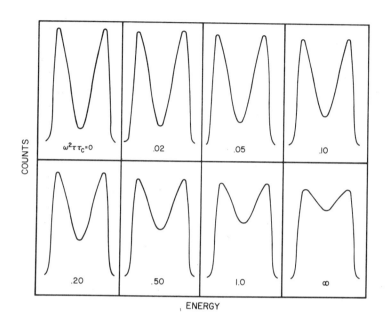

Fig. 12. Theoretical gamma ray lineshape for the 0-2-0 case with perturbation given by the Abragam-Pound theory for time dependent magnetic interactions. Geometrical attenuation factors of $Q_2 = 0.92$ and $Q_4 = 0.76$ were used and all other parameters were the same as for Fig. 11.

## SUMMARY

An excited nucleus ejected at high velocities from a target into gas experiences a sequence of different charge states and different atomic configurations. The hyperfine interaction is therefore of a fluctuating nature and it has been well described by the theory of Abragam and Pound. All of the observations are consistent with a pure magnetic interaction, although it is possible that the effective field is not constant in time. Because of this possibility, g-factor ratio measurements must be performed in such a way that the perturbation is averaged over the same length of time

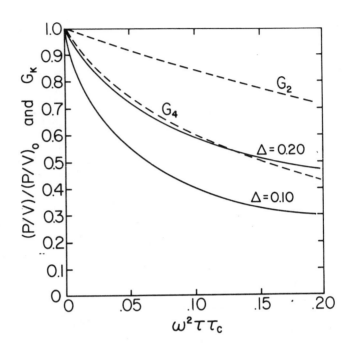

Fig. 13. The attenuation factors $G_2$ and $G_4$ from the Abragam-Pound theory are plotted versus the interaction strength, along with the ratio of the perturbed to unperturbed peak to valley ratios. The parameter $\Delta$ is the ratio of the detector resolution to the total spread in energy of the lineshape. Geometrical attenuation coefficients $Q_2 = 0.92$ and $Q_4 = 0.76$ were used in the line shape calculations.

for both states. With recoil stopper techniques, the g-factor of a 1 psec state can be calibrated with that of a long-lived state in the same atom that can be measured by differential PAC or other precision techniques.

It is possible that ions with the same number of electrons have effective hyperfine fields that are simply related. If this is the case, then relative g-factor measurements can be made in different atoms by carefully adjusting the recoil velocity so that on the average the ions are isoelectronic.

New observation techniques may make it possible to measure very small perturbations, and thus to extend the lifetime range of g-factor measurements to lifetimes of less than a picosecond, where there are many interesting nuclear physics problems.

*Acknowledgements*

The results reviewed in this paper have been obtained in collaboration with several persons. P.D. Bond has done the lineshape calculations. T.R. Miller and M. Takeda have carried the main burden of the data taking and analysis, and A. Whittemore has provided assistance in analyzing the velocity dependence data. Discussions with all of these colleagues and S.S. Hanna have contributed significantly to the content of this paper.

REFERENCES

1. GOLDRING, G. *Hyperfine Structure and Nuclear Radiations* (North-Holland, Amsterdam (1968)), p. 640.

2. BEN-ZVI, I., GILAD, P., GOLDBERG, M., GOLDRING, G., SCHWARZSCHILD, A., SPRINZAK, A. and VAGER, Z. *Nucl. Phys.* A121, 592 (1968).

3. BEN-ZVI, I., GILAD, P., GOLDBERG, M.B., GOLDRING, G., SPEIDEL, K.-H. and SPRINZAK, A. *Nucl. Phys.* A151, 401 (1970).

4. KALVIUS, G.M., SPROUSE, G.D. and HANNA, S.S. *Hyperfine Structure and Nuclear Radiations* (North-Holland, Amsterdam (1968)), p. 686.

5. MILLER, T.R., SPROUSE, G.D., TAKEDA, M. and HANNA, S.S. *Bull. Am. Phys. Soc.* **14**, 1172 (1969) and MILLER, T.R., SPROUSE, G.D., BOND, P.D., TAKEDA, M., LITTLE, W.A. and HANNA, S.S., contribution to this conference.

6. SPROUSE, G.D., HANNA, S.S. and KALVIUS, G.M. *Phys. Rev. Lett.* **23**, 1014 (1969).

7. BRENN, R., LEHMANN, L. and SPEHL, H., contribution to this conference.

8. POLGA, T., RONEY, W.M., KUGEL, H.W. and BORCHERS, R.R., contribution to this conference.

9. DE BOER, J., ROGERS, J.D. and STEADMAN, S., contribution to this conference.

10. FAESSLER, M., POVH, B. and SCHWALM, D., contribution to this conference.

11. GRAHAM, R.L., WARD, D. and GEIGER, J.S. *Bull. Am. Phys. Soc.* **15**, 805 (1970), and private communication.

12. NORDHAGEN, R., GOLDRING, G., DIAMOND, R.M., NAKAI, K. and STEPHENS, F.S. *Nucl. Phys.* **A142**, 577 (1970).

13. ARMBRUSTER, R., DAR, Y., GERBER, J. and VIVIEN, J.P., contribution to this conference.

14. BETZ, H. and SCHMELZER, C. Unilac Report No. 1 (1967) Heidelberg, and BETZ, H. et al., *Phys. Lett.* **22**, 643 (1966).

15. MOAK, C.D., LUTZ, H.O., BRIDWELL, L.B., NORTHCLIFFE, L.C. and DATZ, S. *Phys. Rev.* **176**, 427 (1968).

16. LEPINE, A. and SALAH, O., private communication.

17. BLECK, J., HAAG, D.W. and RIBBE, W. *Z. Physik*, **233**, 65 (1970).

18. LEISI, H.J. *Phys. Rev. A1*, 1162 (1970).

19. BLUME, M. *Hyperfine Structure and Nuclear Radiations* (North-Holland, Amsterdam (1968)), p. 911.

20. BLUME, M. *Phys. Rev. 174*, 351 (1968).

21. GABRIEL, H. *Phys. Lett. 32A*, 202 (1970), and GABRIEL, H. and BOSSE, J. Int. Conf. on Angular Correlations, Delft 1970.

22. BETZ, H.D. and GRODZINS, L. *Phys. Rev. Lett. 25*, 211 (1970).

23. NORTHCLIFFE, L.C. and SCHILLING, R.F. *Nuclear Data A7*, 233 (1970).

24. BHATTACHERJEE, S., BOWMAN, J.D. and KAUFMANN, E.N. *Phys. Rev. Lett. 18*, 223 (1967).

25. BHATTACHERJEE, S., DEVARE, H.G., JAIN, H.C., JOSHI, M.C. and BABA, C.V.K., contribution to this conference.

26. BOND, P.D., SPROUSE, G.D., MILLER, T.R. and HANNA, S.S., contribution to this conference.

DISCUSSION

H.J. LEISI: I would like to bring up the problem of the validity of the Abragam and Pound theory. I think it is rather important and I would welcome all comments concerning this. According to what Dr. Sprouse said, one condition would be that the theory is valid for J much larger than I. This is the limit in which you can replace the quantum mechanical coupling by an effective magnetic field. This is an assumption, and I think it is rather difficult to deduce from this theory statements such as, for instance, that there is no quadrupole interaction present. It is, of course, even more difficult to extract magnetic moments from these measurements.

I would further like to point out that in a recent work (H.J. Leisi, *Phys. Rev. 1A*, 1654, 1662 (1970)) which was mentioned by Dr. Sprouse we proposed a simplified model for explaining, or treating, certain electronic properties of highly ionized ions which we applied to K-capture decays of xenon isotopes. There one finds that the quadrupole interaction should be of the same order of magnitude as the magnetic interaction. This is actually similar to what one finds from atomic beam measurements on rare earths.

G.D. SPROUSE: I think that in the gas measurements there were very large angular momenta involved after the collisions, so it is felt that the J values are quite large. This is not necessarily the case in the vacuum measurements.

H. BERNAS: I would like to go in the same direction as Prof. Leisi. The first thing is that the angular correlation attenuation coefficient is sensitive to the spin correlation function, $<S(t)S(0)>$ (ensemble average). And this, in the Abragam-Pound theory as well as in other theories (say Gabriel or Blume) is usually approximated, just as in NMR or EPR, by a correlation function, that is a product of some correlation amplitude depending on $<\omega^2>$, multiplied by a function of time that is an exponential function, a gaussian function, or something else, according to the person who is making the approximation, and a single correlation time usually comes into this function. Now, the fact that this is an approximation should perhaps be a little more emphasized than many people do in analyzing their data. In particular in solids it is to some degree a very crude approximation. This was emphasized very clearly by Dr. Wickman yesterday. According to the nature of the system you are studying, you may have different correlation times, and you may also, as Dr. Sprouse stated in his talk, have according to the system you are looking at an $<\omega^2>$ which is, in fact, a very complicated average over very large variations and very small variations.

To test the various theories it would be nice to use a system that has a single, spherical, correlation

FAST RECOIL NUCLEI IN GAS    957

time to make analysis simpler. Using recoils in gas is
a good way of doing this, and perhaps the clearest experiment that has been done was that Prof. Leisi reported at
Asilomar, using thermal velocities where you have no
problem of trying to imagine what the correlation time
can be; you can just calculate it from thermodynamics,
using data that is well known in the literature. In this
case, varying the pressure of the gas, you get a whole
range of values $\tau_c/\tau_N$, and you can study the approxi-
mations you are making, and therefore test the various
hyperfine interaction relaxation theories, in a rather
precise way.

After all this, what I would like to know is, or
perhaps what we would all like to know is: What are we
really measuring? When we say we obtain a correlation
time, I think that in a gas with thermal velocities we
know that we are measuring - the spherical correlation
time $\tau_C$ is a good approximation. But then, to interpret
the data of Leisi, we have to introduce a variation of
$<\omega^2>$, and then we are measuring that parameter. When we
are measuring in solids or perhaps - I do not know
whether it is worse or better - in vacuum, $<\omega^2>$ and $\tau_c$
are probably very complicated functions. And I am not
sure that the significance of what we are measuring is
ever going to be very clear. Deducing hyperfine fields
from that is going to be very difficult and deducing
g-factors using those hyperfine fields is perhaps also
going to be difficult.

G.D. SPROUSE: Well, I think I am a little more optimistic
than you in extracting g-factors. The fact that in gas,
$G_2(\infty)$ and $G_4(\infty)$ fall directly on the Abragam-Pound curve,
and that, under the assumption that $1/\tau_c$ is proportional
to the pressure, one observes the pressure dependence to
follow the Abragam and Pound theory, seems to indicate
that things are O.K. So I think that g-factor ratios
measured in gas, if we experimentally make the observation
time the same for two different states, can be very
accurately measured.

H. SPEHL: I would just like to comment on your assumption
No. 3. You have written down that the correlation time

$\tau_c$ has to be small compared to the nuclear lifetime: $\tau_c \ll \tau$. I think the nuclear lifetime $\tau$ in this connection does not have a very physical meaning. You should perhaps use the observed relaxation time instead of the nuclear lifetime. What I mean is that $\tau$ can be very large, but what you have to have is many processes during that time so that the anisotropy is completely wiped out. This applies to the Abragam and Pound as well as to the Dillenburg and Maris expressions.

G.D. SPROUSE: You mean $\tau_c \ll t$, where t is the time of flight, i.e., the observation time?

H. SPEHL: Not the observation time. In the cases where the Abragam and Pound theory applies, the condition is $\tau_c \ll 1/\lambda_k$ where $\lambda_k$ is the relaxation parameter. You have to have many processes during the time the m-substate population goes to uniform population. This time is, in the vacuum case, of the order of 100 psec, whereas the nuclear lifetime can be, of course, 10 μsec.

G.D. SPROUSE: The condition $\tau_c \ll 1/\lambda_k$ is equivalent to $\omega\tau_c \ll 1$. In addition to this, we must have $\tau_c$ less than the observation time.

O. KLEPPER: Is it correct to replace $\omega_Q$ by $\omega_Q \tau_c$ in assumption No. 4? In this case the condition would be fulfilled more easily in the slowing down process of recoiling ions in a gas.

G.D. SPROUSE: Yes, I believe that would be O.K. too. I should put a $t_c$ inside the average, but one must always satisfy the condition that $\omega\tau_c \ll 1$.

D. SPANJAARD: I want to comment on what Dr. Bernas has said. In the Abragam and Pound calculation, the use of one correlation time $\tau_c$ is not fundamental and comes from a particular choice of the electronic correlation function. Therefore one can use another correlation function if one wants to (see D. Spanjaard and F. Hartmann-Boutron, *Journal de Physique*, December 1969).

J.D. BOWMAN: I would like to make an observation about

FAST RECOIL NUCLEI IN GAS                                             959

the assumption that $\omega\tau_c \ll 1$ in the Abragam and Pound
theory used in fitting attenuation parameters $\lambda_2$ and $\lambda_4$
as functions of the pressure of the gas into which the
ion recoils. You will recall that the curve on these
slides deviates from linearity in the low pressure
region. Now, in the low pressure region $\tau_c$ is large
since the collisions are infrequent. You would expect
the actual attenuation to be less than that predicted
from the Abragam and Pound theory. The actual attenuation should tend towards a hard core.

G.D. SPROUSE: Yes, we have considered this explanation
also. When I talk about gas measurements I always mean
staying away from the vacuum region where the curve is
not linear. There was one comment I neglected to make
and that is the following. Another alternative explanation for this deviation at long times is that the ionic
population is ageing. Maybe we can assume that the
Abragam-Pound theory is O.K. in vacuum, it is just that
assumption No. 4 is not correct. You see, here the
Abragam-Pound theory with a constant average value of $\omega$
keeps on going down, but maybe after the nuclei have been
ejected from the foil and have travelled 50 psec, the
field is just turned off. Well, that would cause the
experimental point to be higher.

The simple model of Blume does a good job in fitting
this behaviour, since it mocks this decrease in
perturbation by a pathological hard core which is in
itself entirely unrealistic. So I think that there is
still much work to be done to understand the situation in
vacuum.

S.S. HANNA: I would like to take the privilege of the
chairman to make the last remark, and perhaps try to end
on an optimistic note. I would like to emphasize, I
think, the agreement obtained here for two g-factor
ratios measured by two different methods, the recoil into
gas method and the implantation measurements of
Bhattacherjee. We had been much disturbed that our
measurements at Stanford for silver and rhodium had disagreed with the impact measurements of Wisconsin. The
disagreement in the ratio tabulated by Sprouse comes from

the g-factor of one of the states; the other state seems to agree quite nicely.

Now, I think it is significant that the first state is the short-lived state, and I believe there is a large transient field correction in it in the Wisconsin work. However, in the work of Bhattacherjee, the conditions were such that he did not have to use a transient field correction. Dr. Sprouse was rather careful not to write down any g-factor on the silver and rhodium states, but one can do that if you want to make calculations with known fields adjacent to silver and rhodium; and if you do so, then the g-factors of the Stanford group agree rather well with the work of the Tata Institute.

So I think that the two methods may be coming together.

R.R. BORCHERS: I only wanted to point out that the g-factor measured has extremely large error bars on it. This reflects the fact that we are seeing cancellation of the static and transient fields, as it is probably the field and not the g-factor which is zero.

# TIME DEPENDENT ANGULAR CORRELATION MEASUREMENTS FOR THE FIRST 2+ STATE OF $^{150}$Sm RECOILING INTO VACUUM*

T. POLGA[†], W.M. RONEY, H.W. KUGEL and R.R. BORCHERS

*University of Wisconsin, Madison, U.S.A.*

The time dependence of the angular correlation for the first 2+ state of $^{150}$Sm recoiling into vacuum has been determined using the recoil distance method. γ-rays were detected by four 3 × 3 in. NaI(Tl) crystals in coincidence with backscattered 35 MeV $^{16}$O ions. The target was a 100 μgm/cm$^2$ layer of $^{150}$Sm vacuum evaporated on the reverse side of a 0.25 μm Ni foil. The recoiling ions (with a velocity of 0.013c) were stopped in a 25 μm Cu foil. Both the target and stopper were stretched flat in an assembly designed to minimize γ-ray attenuation. Stopper-to-target distance could be varied by a direct drive micrometer mechanism with an accuracy of ∿ 2.5 μm. An angular correlation for $^{150}$Sm evaporated on the front of a Cu target was measured to determine the zero distance correlation coefficients and to check for differential attenuation in the apparatus. The $A_2$ = 0.648 ± 0.007 and $A_4$ = -1.183 ± 0.011 agree with the predicted values. The mean lifetime of the first 2+ state of $^{150}$Sm and the t = 0 point of the apparatus were determined using the normal recoil distance technique with a high resolution Ge(Li) detector at zero degrees. The average attenuation coefficients ($G_2$ and $G_4$) are shown in Fig. 1 as a function of time-in-flight. The curves in Fig. 1 are calculated

---

*Work supported in part by U.S. Atomic Energy Commission.

[†] On leave from University of Sao Paulo, Brazil

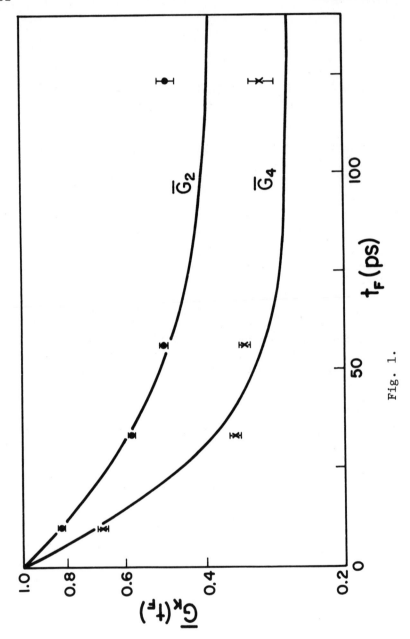

Fig. 1.

assuming an exponential decay $G_k(t) = \exp(-\lambda_k t)$.
The ratio of $\lambda_2/\lambda_4$ determined from the curves is 0.57.
This can be compared to the prediction of the Abragam and Pound theory [1] of 0.3 for a pure time dependent magnetic dipole interaction and 1.7 for a pure time dependent electric quadrupole interaction.

REFERENCES

1. ABRAGAM, A. and POUND, R.V., *Phys. Rev. 92*, 943 (1953).

# VELOCITY DEPENDENCE OF THE ATTENUATION MECHANISM FOR HIGHLY IONIZED IONS RECOILING INTO VACUUM*

J. de BOER, J.D. ROGERS and S. STEADMAN

*Rutgers, the State University, New Brunswick, N.J.*

Measurements of angular distributions of gamma-rays from Coulomb excited $^{114}$Cd nuclei recoiling into vacuum have been made as a function of velocity of the recoil ion. $^{16}$O and $^{32}$S ions of energies between 28 and 75 MeV from the Bell-Rutgers Tandem Van de Graaf were used to bombard thin foils of $^{114}$Cd. The Cd recoil velocity was defined by detecting the backscattered heavy ions in a ring counter, and the coincident gammas were detected in an array of four 3" × 3" NaI crystals. The resulting angular distributions were compared to first order C.E. theory to derive values of the attenuation coefficients $G_2$ and $G_4$ as a function of recoil velocity. The results are shown in the figure. In a second scale the average charge state of the recoil ion is shown [1]. Previous experiments [2] suggest that the atomic excitation produced is sufficiently high that the magnetic fields and correlation times involved can be described as simple functions of Z and v/c. The present results, however, indicate that atomic shell effects may still be important. In this respect we note that the atomic state corresponding to 12 charges removed, where the dependence on v/c in our curves is greatest, is that of neutral krypton.

---

\* Supported in part by the National Science Foundation.

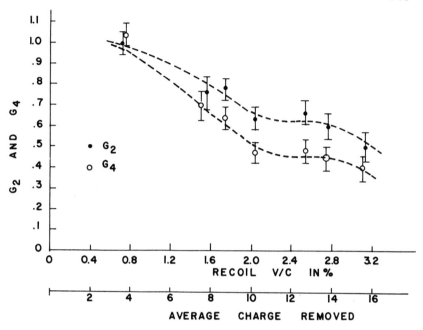

REFERENCES

1. BETZ, H.D. et al., *Phys. Lett.* 22, 643 (1966).

2. BEN-ZVI, I. et al., *Nucl. Phys.* A121, 592 (1968).

# TIME DIFFERENTIAL PERTURBED γ-RAY ANGULAR DISTRIBUTION FROM Yb-NUCLEI RECOILING INTO VACUUM

R. BRENN, L. LEHMANN and H. SPEHL

*Physikalisches Institut der Universität, Freiburg, Germany*

The attenuation of the γ-ray angular distribution following recoil into vacuum after Coulomb excitation of Yb $2^+$ levels with 16 MeV Ne ions has been observed. A kind of time-of-flight technique has been used. The anisotropy is completely destroyed within a time interval of about 200 psec due to time-dependent hyperfine interaction of the nucleus with the ionized and excited atomic surrounding. Experimental results are given in the figure. Least-squares fits to the experimental data have been carried out using two theoretical models. Fit (a) uses the well known Abragam and Pound theory for statistically fluctuating magnetic dipole interaction, predicting attenuation coefficients of the form $G_k(t) = \exp(-\lambda_k t)$. One has to bear in mind, however, that this theory uses first-order matrix elements $Q_{m \to m'}$ for transitions between m-sub-states of the excited nuclear level. Due to the rather strong interaction this is probably too feeble a basis. Though second-order perturbation theory has not yet been carried through, the arguments of Dillenburg and Maris may be helpful. These authors predict, for statistically fluctuating interactions of arbitrary strength, attenuation coefficients of the form $G_2(t) = (1 - \kappa_2) \exp(-\lambda_a t) + \kappa_2 \exp(-\lambda_b t)$, the matrix elements $Q_{m \to m'}$ in this treatment being considered as fit parameters. This procedure has been carried through in fit (b). As can be seen the fit is slightly improved, $\chi^2$ decreasing from 1.55 for fit (a) to 1.13 for fit (b).

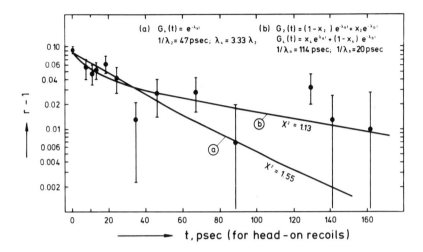

Fig. 1. Attenuation of the γ-ray angular distribution of Yb-nuclei recoiling into vacuum. The measured quantity r - 1, a double ratio of counting rates in two γ-counters at fixed angles, can be thought of as being roughly proportional to $G_2(t)$. The somewhat idealized time scale is given for head-on recoils emerging from the target with recoil energy $E_r$ = 6 MeV. Two least squares fits using different theoretical expressions are shown.

Furthermore, time-intergrated $G_2$ and $G_4$ parameters have been calculated from the fit (b) parameters $Q_{m \to m'}$, and are found to be in good agreement also with the time-integrated experimental data of the group at the Weizmann Institute. It should be pointed out that in both cases the basic assumption is made that the interaction is of a perfectly random type. Any deviations from this assumption would result in a more complicated time pattern of the attenuation parameters.

For a full discussion of the time-of-flight technique, the experimental results and their interpretation, see: R. Brenn, L. Lehmann and H. Spehl, paper submitted to *Nuclear Physics*.

# DEORIENTATION MEASUREMENT IN $^{20}$Ne

M. FAESSLER, B. POVH and D. SCHWALM

*I. Physikalisches Institut der Universität Heidelberg and
Max-Planck-Institut für Kernphysik, Heidelberg*

The velocity dependence of the deorientation effect was investigated for $^{20}$Ne-ions recoiling into vacuum. The reaction used was $^{12}$C($^{12}$C,$\alpha$)$^{20}$Ne$^*$ and for $\alpha$-particles detected under 0° or 180° the $\alpha$-$\gamma$-angular correlation for the first excited state of $^{20}$Ne (E = 1.63 MeV, $I^\pi$ = 2$^+$, $\tau$ = 1.2 psec) was measured. By comparing the perturbed angular correlation (Fig. 1) - given by W($\theta$) = $1 + a_2 G_2 P_2(\cos\theta) + a_4 G_4 P_4(\cos\theta)$ with $\theta$ = $\gamma$-angle in the C.M. system - to the unperturbed one ($G_2 = G_4 = 1$), the attenuation factors $G_2$ and $G_4$ were determined.

The measured attenuation factors (1-$G_4$) are shown in Fig. 2 as a function of the velocity v/c of the Ne-ions [1-$G_2 \approx 0.3(1-G_4)$]. Their magnitude and velocity dependence can be explained by assuming a static, magnetic interaction between the nucleus and its surrounding electrons. Curve a was calculated assuming the magnetic field to be caused by the 1s-electrons of the Ne$^{9+}$ ions only (H = 170 MGauss, g = 0.5). In curves b (g = 0.5) and c (g = 0.7) the contributions of other possible electron configurations were taken into account, using some simplifying assumptions.

# DEORIENTATION MEASUREMENT IN $^{20}$Ne

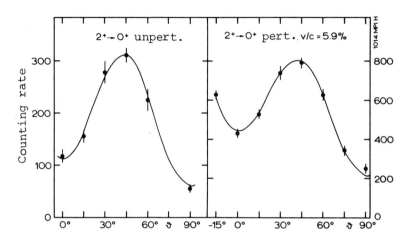

Fig. 1. α-γ-angular correlation.
Right: perturbed correlation, recoil into vacuum, $v/c(^{20}Ne) = 5.8\%$.
Left: unperturbed correlation measured at the same beam energy; recoil into a gold backing.

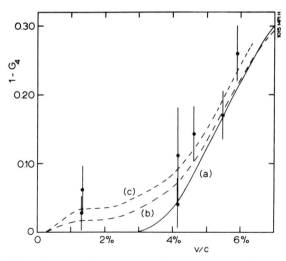

Fig. 2. Measured attenuation factors $1-G_4$. The curves are discussed in the text.

# DETERMINATION OF CROSS SECTION FOR CHARGE EXCHANGE AND DEPOLARIZATION IN GASES FROM PERTURBED ANGULAR CORRELATIONS

F.N. GYGAX and H.J. LEISI

*Laboratory for High Energy Physics, Swiss Federal Institute of Technology Zurich, Switzerland*


Angular correlations measured on paramagnetic ions in a gas were found to exhibit a characteristic drop of the anisotropy as the pressure of a buffer gas is increased, starting from the region of pressure-independent attenuation [1]. The effect is induced by atomic collisions and results from a change of the hyperfine coupling during the lifetime $\tau$ of the intermediate state, either by a change of the atomic state (e.g. by charge exchange) or by electronic spin depolarization. The corresponding cross section can be determined from such experiments.

We have measured the anisotropy of the 172-203 keV $\gamma$-$\gamma$ angular correlation in the decay of gaseous $^{127}$Xe as a function of the density of iodine buffer gas (see Fig. 1). The data can be analyzed in terms of average attenuation factors $G_2^0$ and $G_2^1$ which correspond to decays having suffered no collision [2], and one or more collisions, respectively. The measured attenuation factor is approximatively given by

$$G_2 = (1 - P)G_2^0 + PG_2^1, \qquad (1)$$

where

$$P = \exp\{-\tau_c/\tau\} \qquad (2)$$

is the fraction of decays with one or more collisions,

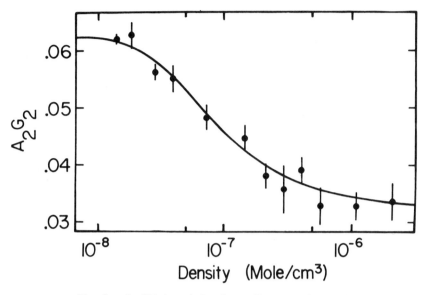

Fig. 1. Coefficient $A_2G_2$ of γ-γ directional correlation as a function of the density of iodine. Solid line: fit based on Eq. (1).

$\tau_c$ being the mean time before the first collision occurs. The time $\tau_c$ is inversely proportional to the density and can be expressed in terms of a cross section. The results of this analysis for the present experiment and the earlier experiment with xenon buffer gas [1] are shown in Table I. The measured cross sections are one and two orders of magnitude larger than the geometrical cross sections.

We have attempted to explain these unusually large values by invoking a classical collision model. The electric field produced by the highly charged iodine ion (charge Ze) polarizes the atoms of the buffer gas. An attractive potential causes important deviations from a straight flight pass and produces geometrical collisions for impact parameters which are smaller than a critical value

$$b_{crit} = \left\{ \left(\frac{\alpha}{2M}\right)^{1/2} \frac{4Ze}{v} \right\}^{1/2} \tag{3}$$

Here, $\alpha$ is the polarizability of the atoms of the buffer gas, M is the atomic mass (assumed to be the same for ions and buffer gas atoms), and v is the mean relative velocity of the two colliding bodies. The cross section deduced from $b_{crit}$, Eq. (3), is listed in the last column of Table I for the case of xenon buffer gas. The proposed mechanism accounts for a large value of the cross section which is nearly in agreement with the experiment, at least in the case of xenon buffer gas.

TABLE I. Cross Section for Collisions between Iodine Ions (Average Charge Seven [3]) and Different Buffer Gases.

| Buffer gas | $\sigma[10^{-15} \text{ cm}^2]$ | | |
|---|---|---|---|
| | experimental*) | geometrical | theoretical |
| Xe | 90 ± 45 | 7 | 45 |
| $I_2$ | 800 ± 400 | 11 | – |

*) preliminary evaluation

REFERENCES

1. GYGAX, F.N., EGGER, J. and LEISI, H.J. *Proceedings of the Conference on Electron Capture and Higher-Order Processes in Nuclear Decays*, Debrecen, Hungary, July 1968, edited by D. Berényi (Eötvös Lòrànd Physical Society, Budapest, 1968), 386.

2. LEISI, H.J., *Phys. Rev. 1A,* 1654 (1970).

3. LEISI, H.J., *Phys. Rev. 1A,* 1662, (1970).

# POSSIBLE TDPAC FOLLOWING LOW-ENERGY RECOIL INTO VACUUM

R. ARMBRUSTER, Y. DAR*, J. GERBER and J.P. VIVIEN

*Centre de Recherches Nucléaires, Strasbourg, France*


Measurements by Ben Zvi et al. [1] indicated perturbations in the angular distribution pattern of the radiation emitted by excited nuclei recoiling into vacuum. The distributions are attenuated due to hyperfine interactions with the atomic shells.

The temporal behaviour of such an attenuation was measured by means of a genuine time-differential method by Brenn and Spehl [2]. The Coulomb excited rare-earth nuclei recoiled into vacuum with energies as high as 3 MeV in the forward direction. The gross features of the measurement show that the interaction results in a washout of the distribution in about 200 ps in a roughly exponential behaviour, with a time constant 80 ps.

In the following a time differential measurement is described, where recoil energies are an order of magnitude lower than the previously mentioned work. The present work, which serves as a test case both for a perturbation and for the experimental implications, has a complementary part, namely integral measurements. The present contribution reports on the time differential part. Following Coulomb Excitation, the angular distribution of K-conversion electrons can be written [3] as:

---

*Present address: Department of Physics, Technion, Haifa, Israel

$$W(\theta,t) = 1 + \sum_{2,4} Q_k G_k(t)\, a_k(\xi)\, A_k B_k^K P_k (\cos\theta)$$

where: $Q_k$ = geometrical attenuation coefficients, $G_k$ = attenuation coefficients, $a_k$ = Coulomb Excitation parameters, $A_k$ = gamma directional correlation coefficients and $B_k^K$ = K conversion particle parameters.

With an available beam of 4 MeV alpha particles, for almost every low-lying level in a rare earth nucleus, $a_4$ vanishes [3]. One expects, under such conditions, to deduce information on $G_2$.

PAC time-dependent effects can be observed especially when the characteristic times involved differ considerably from the nuclear lifetime and therefore modify the decay pattern. As a check on the experimental set-up, the isotropic radiation from the 9/2 level in $^{177}$Hf was observed, exhibiting a pure nuclear decay. It would now suffice in principle to observe any deviations from a normal time distribution.

The measurements were carried out with the microwave method [4,5] for determining lifetimes of nuclear excited states.

The $2^+$ level, 122 keV in $^{152}$Sm, was chosen for the first measurement mainly due to the requirements of the microwave method, which utilizes conversion electrons. Two targets of $Sm_2O_3$ were evaporated, so that there was a considerable difference in the amount of the two fractions that recoiled into vacuum out of the targets. Low statistics prevent using too thin a target for a large recoil-into-vacuum fraction, and electron line width requirements set an upper limit for the thickness. In addition, the sharp dependence of the stopping power for the recoiling nuclei in the low-energy region prevents accurate calculations for optimum target thicknesses [6].

Targets of 30 µg/cm² and 100 µg/cm² were desired. Those available were 40 µg/cm² and 80 µg/cm². While the ratio is known quite accurately, the absolute values may be lower by 25% due to inconsistency between supplied gold standards.

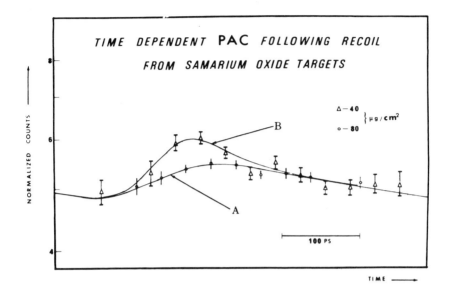

Fig. 1. Time spectra of the radiation from the $2^+$ level in $^{152}$Sm as observed with the microwave method [4,5]. The experiments were performed with two targets differing in thickness. The circular points were taken with a target of 80 µg/cm$^2$ and fit curve A very well. A represents the calculated spectrum for a decay characterized by the nuclear lifetime. The triangular points taken with a target of 40 µg/cm$^2$ deviate from A. B is the calculated behaviour when a component with 35 ps is present in the radiation. Due to finite resolving times one cannot exclude $\tau_2 \leq$ 90 ps. The two distributions were normalized at the delayed end of the time region.

The measurements indicate that for the thicker target a normal behaviour is observed corresponding to the nuclear decay (see [4] or [5]). This does not exclude, in principle, effects with characteristic times much longer than 1 ns, for example, which cannot be observed with the present apparatus.

With the thin target, however, one observes a small component, of about 1%, characterized by a fast decay. Assuming

$$G_2 \sim e^{-t/\tau_2}$$

we get the best fit for $\tau_2 = 35$ ps. In this time region the fits are very flat, and therefore one should set only a limit. A detailed analysis cannot exclude $\tau_2 \leq 90$ ps (compared with [2]).

Under the experimental conditions

$$W(\theta,t) = 1 + 0.12\, e^{-t/\tau_2}$$

If the fast component stems from recoiling into vacuum, it means that a minimum 10% of the total is perturbed. This and the fast $\tau_2$ somewhat contradict present knowledge.

## REFERENCES

1. BEN ZVI et al., *Nucl. Phys.* *A121*, 592 (1968).

2. BRENN, A. and SPEHL, H., Abstract submitted to the Heidelberg Conference, 1969.

3. ALDER, K. et al., *Rev. Mod. Phys.* *28*, 432 (1956).

4. BLAUGRUND, A.E. et al., *Phys. Rev.* *120*, 1328 (1960).

5. ARMBRUSTER, R. et al., *Nucl. Phys.* *A143*, 315 (1970).

6. LINDHARD, J. et al., *Mat. Fys. Medd. Dan.* *33*, No. 14.

# MAGNETIC MOMENT DETERMINATION AT PRE-SET TRAVEL TIMES IN THE RECOIL-INTO-GAS METHOD*

T.R. MILLER, G.D. SPROUSE, P.D. BOND, M. TAKEDA,
W.A. LITTLE and S.S. HANNA

*Physics Department, Stanford University,
Stanford, Calif., U.S.A.*

Two techniques have been used in measuring magnetic moments by the recoil-into-gas method. In one, the recoiling ions are allowed to travel for times long compared to the lifetimes of the nuclear state [1]. In the other, the travel times are restricted to times less than or comparable to the lifetimes by means of a stopper placed at a desired distance [2]. The latter technique has the advantage that the travel time can be controlled so that a possible variation of the hyperfine interaction with time can be studied and taken into account in the evaluation of the magnetic moment. In this communication we apply this technique to levels with lifetimes in the picosecond range [3]. The principal aims are to improve the analysis and precision of the magnetic moment measurements for these very short lived nuclear states, as well as to study the dynamics of hyperfine fields in recoiling ions.

The technique was applied to the two levels in $^{109}$Ag at $E_x$ = 414 keV, $\tau$ = 55 psec and $E_x$ = 309 keV, $\tau$ = 9 psec. In the earlier work [3] the travel time of the recoiling ions was unrestricted. In the present experiment, the average travel time for the Ag recoils, excited by 52 MeV

---

*Supported in part by the U.S. Army Research Office (Durham) and the National Science Foundation.

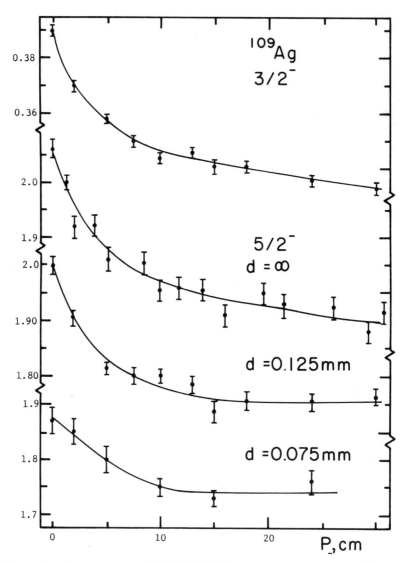

Fig. 1. Top curve: W(0°)/W(90°) vs. pressure for 3/2⁻, $\tau$ = 9 psec state of $^{109}$Ag at stopper distance d = 0.125 mm. Bottom three curves: W(90°)/W(0°) vs. pressure for 5/2⁻, $\tau$ = 55 psec state at the distances indicated.

$^{35}$Cl ions, was restricted to times comparable to or shorter than the lifetimes by means of a copper stopper placed after the target. In all other respects the method, which involves "singles" counting, was the same as that used previously [3]. The perturbation of the gamma-ray correlation was determined by measuring the 90°/0° ratio of the yields with two fixed Ge(Li) detectors. Fig. 1 shows the ratio vs. pressure (He gas) curves for the 3/2$^-$, 9 psec state at a stopper distance of 0.125 mm and for the 5/2$^-$, 55 psec state at distances of $\infty$, 0.125 and 0.075 mm. The latter distances correspond to average times of 20 and 12 psec. Inspection of these curves clearly shows the effect of shortening the travel time. At the shorter distances the perturbation in vacuum is reduced and restoration of the correlation occurs at lower pressures.

The curves in Fig. 1 have been fitted with the theory of Abragam and Pound [1] and values of $\omega\tau$ have been extracted. These values are averages over the velocity distribution and the flight paths of the recoils. To obtain the quantity $\omega\tau$ accepted values of the correlation time $\tau_c$ have been used [1]. The derived values of $<\omega\tau>$ are shown in Fig. 2. For the short lived nuclear state the stopper position has very little effect on the value of $<\omega\tau>$, as expected, since the mean life $\tau$ is less than the flight time $t_s$ and most of the nuclei decay in flight. For the long lived nuclear state it appears that $<\omega\tau>$ may increase as $t_s$ becomes much shorter than $\tau$. This would indicate that the hyperfine field is time dependent and decreases with increasing time.

REFERENCES

1. BEN-ZVI, I., GILAD, P., GOLDBERG, M., GOLDRING, G., SCHWARZSHILD, A., SPRINZAK, A and VAGER, Z., *Nucl. Phys. A121*, 592 (1968).

2. SPROUSE, G.D., HANNA, S.S. and KALVIUS, G.M., *Phys. Rev. Lett. 23*, 1014 (1969).

3. MILLER, T.R., SPROUSE, G.D., TAKEDA, M. and HANNA, S.S., *Bull. Am. Phys. Soc. 14*, 1172 (1969).

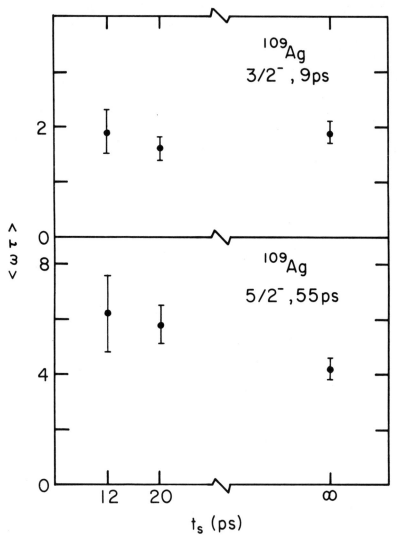

Fig. 2. Values of $\langle\omega\tau\rangle$ derived from the curves of Fig. 1.

# MAGNETIC MOMENT DETERMINATION BY LINE SHAPE ANALYSIS IN THE RECOIL-INTO-GAS METHOD*

P. D. BOND, G. D. SPROUSE, T. R. MILLER and S. S. HANNA

*Physics Department, Stanford University, Stanford, Calif. U.S.A.*

It has been shown [1] that the angular correlation in a reaction produces a characteristic Doppler broadened line shape for an emitted gamma ray. The $0^+ \to 2^+ \to 0^+$ sequence in Coulomb excitation, viewed at 90°, produces an exceptionally exotic shape consisting of a resolved doublet. The valley arises from the fact that for this spin sequence γ-rays are not emitted normal to the scattering plane. Since the shape of this doublet, in particular the depth of the valley, is a very sensitive function of the particle-gamma correlation, it can be used as a measure of any perturbation of the correlation. In the present application the shape is used to determine the perturbation produced by the hyperfine field in highly ionized ions recoiling into gas. As in the usual recoil-into-gas method [2,3] the perturbation is determined as a function of the gas pressure. The advantages of the line shape technique are that it involves "singles" detection at a single angle and gives high sensitivity. The chief limitation is the requirement that the detector resolution be sufficiently good to reveal the main structure of the line. At present this limits the method to gamma rays with energies greater than about 0.5 MeV.

The method has been tested on the 0.847-MeV, $2^+$ state

---

*Supported in part by the U.S. Army Research Office (Durham) and the National Science Foundation.

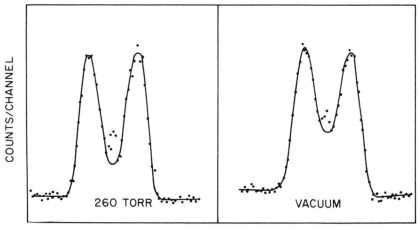

Fig. 1. Line shapes of the 0.847 MeV γ-ray of $^{56}$Fe, Coulomb excited by 60 MeV $^{35}$Cl ions and observed at 90°. Left, $^{56}$Fe ions recoiling into argon gas at 260 Torr pressure; right, into vacuum.

of $^{56}$Fe, τ = 11.0 psec and g = 0.55. A $^{35}$Cl beam was used to Coulomb excite the state at three beam energies: 48, 56 and 60 MeV. Line shapes of the 0.847-MeV γ-ray, detected with a 50 cc Ge(Li) detector at 90°, are shown in Fig. 1 for recoil into vacuum and into argon gas at a pressure of 26 cm Hg. At the latter pressure the angular correlation is essentially unperturbed. In the vacuum run the valley has been partially filled in by the perturbation arising from the hyperfine field. The small peak which appears in the valley comes from recoils which have stopped in the target. Fig. 2 shows two "perturbation vs pressure" curves. In the top curve the perturbation is determined by measuring W(90°)/W(0°) with two fixed detectors. In the bottom curve the perturbation is determined from the line shape: the quantity plotted is the valley/peak ratio of the doublet (Fig. 1). The trend of both curves is consistent with the known values of $H_{int}$ [3] and g for $^{56}$Fe. The significant feature of Fig. 2 is the greatly increased sensitivity of the line shape method (≈2:1 change with pressure) over the 90°/0° ratio method

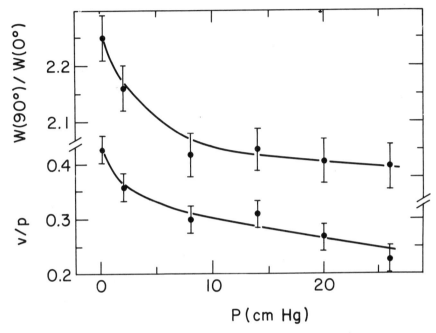

Fig. 2. Perturbation of correlation of 0.847 MeV γ-ray of $^{56}$Fe as a function of gas pressure. Top: W(90°)/W(0°). Bottom: valley-peak ratio of line shape.

(≈10%). The experimentally observed line shapes have been fitted with curves calculated from the theoretical correlation function in order to determine the magnitude of the perturbation.

Line shapes calculated for other spin sequences show that the method can be used effectively in several other cases; the sequence 1/2 → 5/2 → 1/2 is especially attractive.

REFERENCES

1. FISHER, T. R., HANNA, S. S., HEALEY, D. C. and PAUL, P., *Phys. Rev.* 176, 1130 (1968).

2. BEN-ZVI, I., GILAD, P., GOLDBERG, M., GOLDRING, G., SCHWARZSCHILD, A., SPRINZAK, A. and VAGER, Z., *Nucl. Phys. A121*, 592 (1968).

3. SPROUSE, G. D., HANNA, S. S. and KALVIUS, G. M., *Phys. Rev. Lett. 23*, 1014 (1969).